Washington, DC
Oct. 10, 1949

It began with J. Carlton Ward Jr. World War II was over, the aviation industry was about to lose its government contracts, and Ward was on Capital Hill to testify on behalf of those acres of draftsmen with no more bombers to draw. He was the president of Fairchild Engine & Airplane Corp., a tall man with a patrician air and a voice that commanded attention, especially when he digressed from his testimony about procurement policy and happened to mention building an atomic-powered plane. "It was an inadvertent remark," he remembers, a notion inspired by a report on the atomic bomb that he had just finished reading. Senator Homer Ferguson immediately interrupted "You see a future for atomic power in an airplane?" "I think so," replied Ward, going on to predict that it would be just as revolutionary as the jet engine. "The whole tactical concept of war will change to the nation that first solves that problem." "Would you go so far as to say," asked Senator Hugh Mitchell, "that almost any amount of money spent by the government in well-conceived experimental programs in the development of the utilization of atomic energy as a propulsion force would be justified?" "I think our nation can't afford to do otherwise if it wants to be first," said Ward. "In other words," said Ferguson, "we have really got to keep out in front in science?" "If we want to keep out in front in international affairs."

Science 1982

ELEMENTS IN THE BOOK

The Early Years	**3**
Hardware Development	**4**
Flight Test - the NB36	**6**
Design of the Reactor	**10**
Convair X-6	**11**
The X-211 Flight Test Powerplant	**16**
NX-2 A/C	**22/25**
X310	**23**
XNJ140	**24**
Developments during the Latter Years	**34**
The Idaho Test Station	**38**
Russian Developments	**45**
Human Factors	**56**
The End of the Program	**58**
Appendix	**60**
Program Costs	**73**

Dreaming

ALL ENERGY HAS ALWAYS COME FROM "OUT OF THIS WORLD!" THE FAR-OFF SUN HAS GIVEN US INDIRECTLY THE **STORED** ENERGY OF COAL AND OIL...THE **LIVING** ENERGY OF PLANTS AND ANIMALS (AND OF HUMANS, TOO).

TODAY, SCIENTISTS HAVE FOUND THE SOURCE OF THE SUN'S STRANGE AND WONDERFUL ENERGY LOCKED IN THE HEART OF THE ATOM... ARE RELEASING THAT **ATOMIC** ENERGY TO SERVE US ALL IN THE FUTURE AS A SOURCE OF ALMOST UNLIMITED POWER.

By 1948 the dreaded Mushroom cloud and all its associated fear was being replaced by "Nucleomania". As we can see below the uses seemed endless. Could atomic bombs be detonated over the North Pole to warm the ice caps and give the world a more moderate climate? That was a question of the times.

Popular Mechanics Magazine

WRITTEN SO YOU CAN UNDERSTAND IT

Vol. 75 — JANUARY, 1941 — No. 1

The MIRACLE of U-235

By Dr. R. M. Langer

SEVERAL months ago you read in Popular Mechanics that the secret of atomic power is close to solution and that possibly within a decade civilization will be using the tremendous amounts of energy contained in U-235, an isotope of uranium.

Since then, additional research has opened up greater possibilities than ever. At first we thought that a heavy power plant weighing tons would be required to extract the power; now it appears that a pound or even a single ounce of

January 1941

Depictions of nuclear-powered jet flying wings were not uncommon in the popular semi-technical press in the 1940s, bridging the gap between pulp science fiction spaceships and the world of visionary technology. For instance, the January 1941 issue of Popular Mechanics showed this concept for a U-235-powered wing. Other concepts envisioned winglike aircraft propelled by beams of incandescent metal heated by nuclear fission.

Gen. Comics 1948

In its August 25, 1945 issue, Newsweek published this image of a futuristic "atomic-powered" flying-wing. The amount of fuel needed to propel the plane was represented by the circled dot, according to the article.

The Early Years

As early as 1943, American scientists were thinking about the possibility of propelling large aircraft with nuclear- powered jet engines. However, US Government officials advised these scientists to defer such futuristic applications because all efforts during the War were to be directed towards the urgent development of a nuclear bomb.

Interest in atomic energy climbed into high gear after World War II ended. The very scientists who had raced to produce a bomb had also developed theories for a number of possible alternative uses for nuclear energy. These ideas included: electric power generation, nuclear excavation of large land masses, and propulsion systems to power vehicles on the land (trains), sea, and the air. The military were particularly interested in a nuclear-powered aircraft. This is the very story that will unfold in "NX-2".

Soon after the Hiroshima and Nagasaki nuclear explosions in August of 1945, Gordon Simmons, a young engineer involved in the construction of the Oak Ridge plant, prepared a letter to the Fairchild Aircraft Company stating that he would like to be associated with a company interested in designing nuclear- powered aircraft. His thoughts had been stimulated by J. Carlton Ward, president of the Fairchild Co. Mr. Ward had eloquently expressed the strategic advantages to be gained by harnessing nuclear energy to power large aircraft. Before the end of October 1945, Fairchild decided to make presentations to the U.S. military services for sponsorship of such a project. After numerous conferences, the Army Air Corps decided to sponsor a unified project with the National Advisory Committee for Aeronautics (NACA) participating.

In January 1946, the U.S. Air Force (USAF) asked industry members to select one of their companies to be the single manager of the group effort, and prime contractor to the USAF. Fairchild Engine and Airplane Company was selected as the leader. The program would be known as NEPA, Nuclear Energy for the Propulsion of Aircraft. The team of contractors included: Allison, United Aircraft, Wright Aeronautical, General Electric, Westinghouse, Continental Aviation, Lycoming, Northrop, Flader, and Menasco Mfg. Member companies were to have a voice in technical phases of the NEPA Program, and could assign personnel to the working organization. Contracts were signed authorizing funding of $5.25 million. Late in 1946, the NEPA Program office moved to Oak Ridge, Tennessee. That same year, Johns Hopkins University studied the potential as well as the problems of using nuclear power in aircraft. Chief amongst the problems at the time was the lack of data on the effects of radiation on materials in general. Another basic problem was the release of radioactive products during an accident. Also, shielding the crew and ground personnel from high levels of radiation was a serious problem that needed further study.

After considerable thought, the NEPA Program office recommended that the major effort be concentrated on a direct air-cooled ceramic reactor, powering a turbojet engine. The team and Government agencies agreed. Later, the Defense Department became involved and recommended that NEPA proceed on a priority basis as a coordinating project with the Atomic Energy Commission (AEC). In January 1948, the Finletter Report recommended intensifying research efforts on a nuclear plane. Two months later, Congress urged that NEPA be given the highest priority in atomic energy research. Later that year, an MIT report predicted that a nuclear- powered aircraft was feasible, and could be produced within 15 to 20 years, at a cost of $1 + billion. Total NEPA personnel were now up to 444.

By February of 1949, the NEPA Program became a joint effort of the US Air Force, US Navy, the AEC, and NACA. During that year, many studies and breakthroughs were achieved. These included: improved shielding for the crew compartment, improvements in turbojet cycling, fabrication of beryllium bodies, circulating fuel type reactors were studied, and the testing of liquid metals handling was undertaken. Also, a small- scale air cycle powerplant was constructed using a turbojet engine and an electrical heat source. **(It was said by some that the electric heat source experiment was to prove to the skeptics that one did not need to burn fuel to turn turbine blades.)**

This was an attempt to simulate a nuclear reactor. A survey of jet engine manufacturers was made to establish limitations on engine size and characteristics.

In 1950, the concept of a water-moderated reactor was patented. This made operation of a reactor in an aircraft feasible. Also, that year, the AEC created an Aircraft Reactors Branch, absorbing the AEC portion of NEPA. NEPA personnel peaked at 676 (including 195 scientists and engineers). Average salary of engineers with 10 years experience was $475 per month. Starting salary for new hires was $300 per month. That November, Fairchild entered into a contract with the AEC for work related to the nuclear aircraft program. Almost simultaneously, the US Air Force advised Fairchild that the first phase of the Aircraft Nuclear Program was at an end. Recall that the Korean War began in June 1950.

Hardware Development Phase

As January 1951 began, the USAF decided that the program had advanced for work to begin on "hardware development," and the stage of feasibility studies should be closed. The mission of the NEPA Project had been accomplished. The hardware development phase would now be carried out by General Electric Company, and development of the aircraft was placed with Consolidated Vultee Aircraft Company (Convair). During the five- year life of the Fairchild NEPA program, total funding was $24 million. The high quality of NEPA employees resulted in many scientific studies and tests to evaluate possibilities. Most of the best technical people had numerous job offers in the future program.

However, the design of the reactor had not been fully established. Two methods were under consideration: the liquid coolant (sodium) cycle and the direct air cycle.

Starting in May 1951, General Electric initiated an Aircraft Nuclear Propulsion (ANP) in its Aircraft Gas Turbine Division at Evendale, Ohio. Eighty-seven veteran NEPA Program personnel joined GE, and continued to work at Oak Ridge, TN while facilities at Evendale were prepared for the nuclear research. D.R. (Roy) Shoults was selected as the program manager, and M.C. Leverett was the Engineering Manager. The first six months were devoted to the evaluation of the two possible power cycles. By October, the direct air cycle was selected for development, and work on liquid metals was phased out at GE. Also that year, the configuration for the nuclear reactor was successfully fabricated. This consisted of fuel elements of uranium dioxide cores, with a stainless steel cladding. This approach was used throughout the program, including the High Temperature Reactor Experiments later conducted in Idaho. Total personnel in the GEANP was up to 450.

Logo for the "Aircraft Nuclear Propulsion" program

HTRE-1

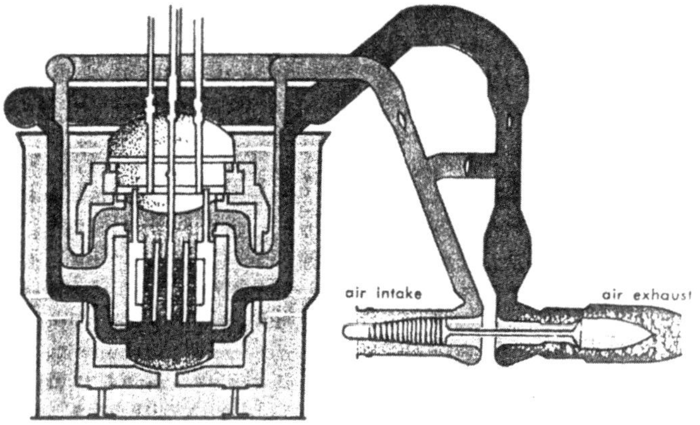

Crossection of the direct cycle testrig.
"High Temperature Reactor Experiment"
Later called "Heat Transfer Reactor Experiment"

In 1952, the Government made two key pronouncements: First, AEC approved the use of the National Reactor Test Station (NRTS) at Arco, Idaho, as the flight test base. Also, the AEC and the Department of Defense determined that plans needed to be made for flight test of a nuclear reactor system in the 1956-1958 time period, utilizing a modified B-36 strategic bomber as the test bed (Non-propulsion system). Program plans were also

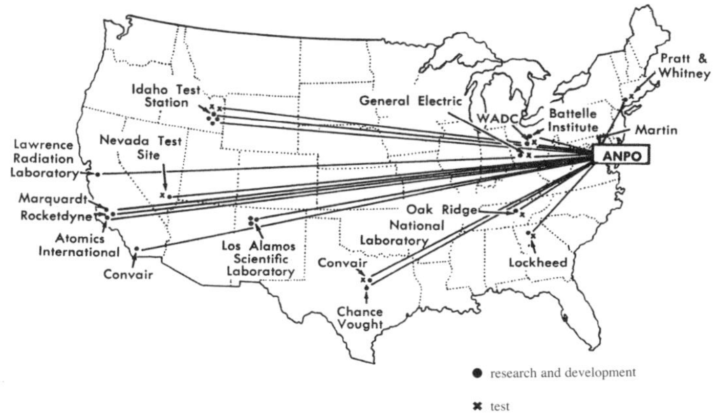

Facilities available to the Aircraft Nuclear Propulsion Office.

made to use a Globemaster cargo aircraft as a full scale test aircraft, beginning in the late 1950's.

In December 1952, the Office for Aircraft Nuclear Program was established to coordinate AEC and USAF participation. Major General Donald Keirn was named director. He was an early champion of nuclear flight. General Keirn first was involved in jet engine work as the coordinator between the USAF and General Electric during development of the first US jet engine. This was the GE1-A engine, which powered the Bell XP-59A Airocomet, which first flew in 1942. During this time, the general became acquainted with Roy Shoults, then an engineering manager with GE. Keirn and Shoults spent time together during the 1940's, discussing linking the jet engine with nuclear power for the propulsion of aircraft.

Early in January 1953, the Eisenhower Administration began. Three months later, the National Security Council ordered the AEC and the Department of Defense (DoD) to cancel the ANP Program on the grounds of budget savings, and that the program was not in the national interest. Secretary of Defense Charles Wilson ordered the program canceled. Mr. Wilson had headed the General Motors Co. before being named as Defense Secretary. Within a month, the USAF and other members of the DoD succeeded in re-opening the project with a major reorganization. The ANP was re-directed toward applied research and development on a limited funds basis. A series of high temperature reactor experiments (HTRE) were scheduled to develop and prove out the reactor power plant. That year, at GE, the ANP became the Aircraft Nuclear Propulsion Department (ANPD). This was established under the GE Atomic Products Division. In December, the USAF informed the AEC of its renewed interest in manned nuclear aircraft, and asked the AEC to expedite experimental work. This political in-fighting would continue throughout the program.

In April 1954, the USAF director, General Keirn, advised the Joint Committee that a nuclear- powered aircraft could be in operation in half the scheduled time if given a high priority. The Joint Committee approved a report calling for a "crash" program. This report was forwarded to President Eisenhower, the DoD secretary, and the AEC chairman. In July, the AEC decided to fund a second development program, with Pratt & Whitney. This was to study the indirect liquid metal cycle propulsion system. During 1954, a GE- developed concentric- ring fuel element design for the reactor core was proven successful. Fabrication techniques were also developed establishing hydrided zirconium as a practical solid moderator material. This design was used in the final GE reactor design. Meanwhile, in Evendale Ohio, critical experiments were being initiated to provide data for design of the first actual reactor, the HTRE-1. Late that year, the USAF outfitted a C-46 transport with passenger seats and other amenities for transporting GE Evendale personnel to the Idaho Test Station. This USAF operated "Site Flight" service required an eight- hour flight from Ohio to Idaho.

HTRE-2 as seen today at Arco, Idaho site. Actually a rebuild of HTRE-1. Total weight over 600,000 lbs.
INEEL photo

By February 1955, the AEC reported that progress on the direct cycle reactor exceeded expectations, and authorized additional funds to be spent in FY 1955. Two months later, the US Air Force issued requirements for a Weapon System 125-A high performance nuclear powered aircraft, and initiated a Program Office at Wright-Patterson in Dayton, Ohio.

In June, the AEC and the DoD agreed to accelerate the ANP Program with the objective of testing a prototype by 1959-1960. In September, Pratt & Whitney was authorized to begin work on the indirect cycle reactor, using liquid metals. Construction of CANEL (Connecticut Aircraft Nuclear Engine Laboratory) was started that year. Progress was much slower on the indirect cycle approach. P&W never ran a practical test system, whereas, the GE reactors and jet engine tests were successful. The P&W work was limited to component testing. In the long run, the indirect cycle showed more promise, but it also required a lot more development work.

In September 1955, the NB-36 bomber with a 1 megawatt reactor had its first test flight. This reactor was not used for propulsion, but to test the effects of a nuclear reactor in a flying aircraft. A total of 47 flights were made in two years from Carswell AFB, Texas. The reactors were powered up over the New Mexico desert.

B-36H

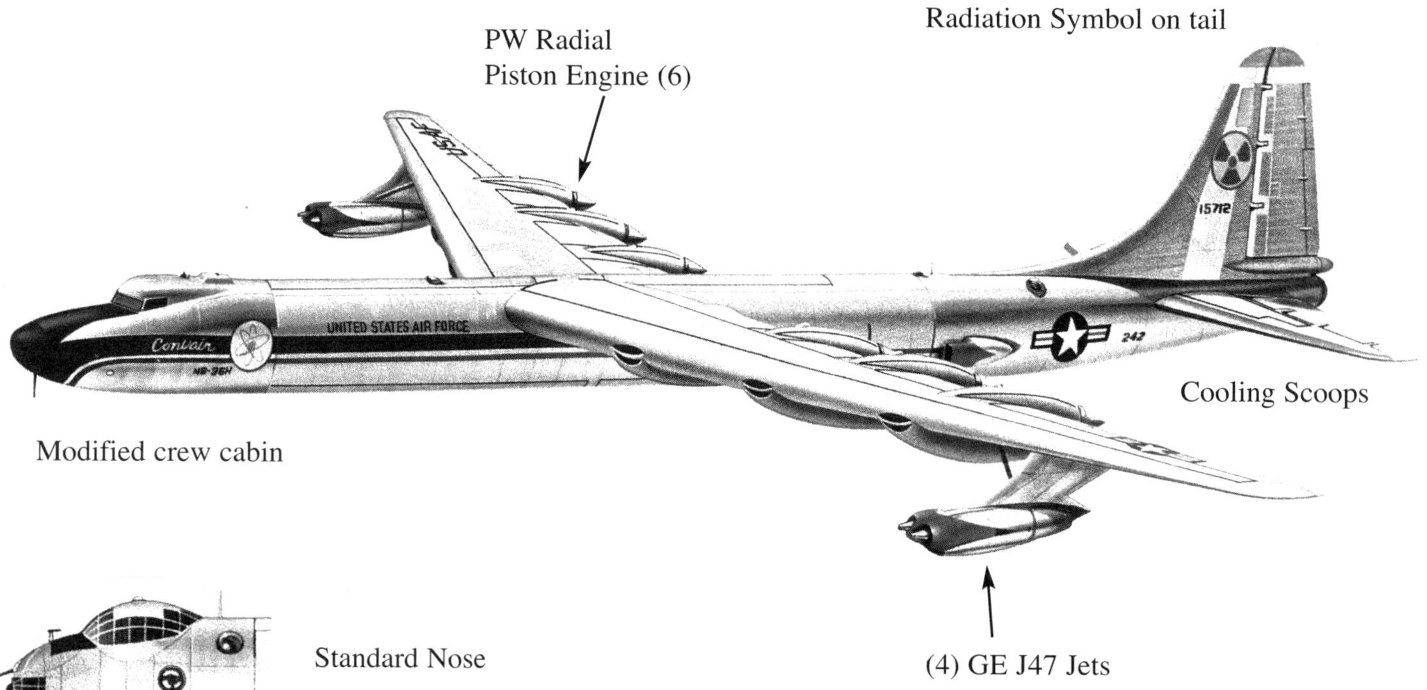

Standard Nose

P1 Test Reactor

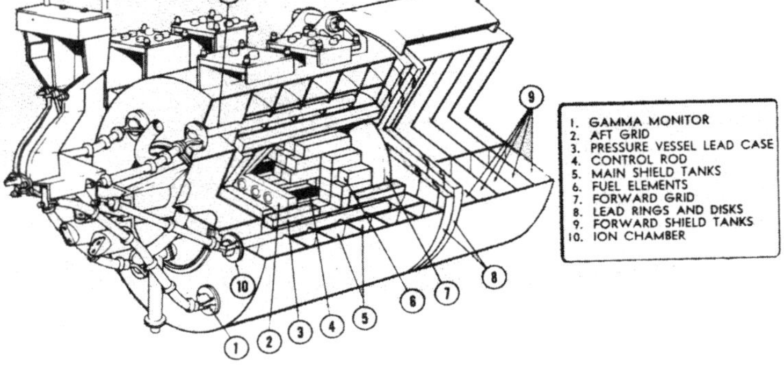

Test reactor, shown in cutaway view above, weighs more than 20 tons. Airborne reactor can be operated at power levels as high as 1,000 kw. at altitudes up to 40,000 ft.

NB-36 Test Program

When it became evident that flight tests would be necessary to get adequate shielding and nuclear processing data, the mighty B-36 was chosen because of its huge fuselage (162 feet long) and great payload capacity. The nose section was removed from a B-36H, and was replaced with a nose that would accommodate the shielded crew compartment.

Air Intake on aft section of modified N-36H leads to heat exchanger section

40,000 pounds of bombs. In these tests, the reactor was not producing any propulsive output. Reactor heat was carried off through heat exchangers, cooled by ram air from a scoop on each side of the fuselage.

Left photo of the B-36H built by Convair Division of General Dynamics Corp. Ft. Worth, Texas. It carried the nuclear test reactors during 47 flights from Carswell AFB, Texas over a two year span. Actual power up of the reactors was accomplished when the aircraft was over the New Mexico Desert.

Localized beefing of the fuselage was adequate to accept the 12 ton crew capsule. The shielded compartment had room for five crew members. Shielding consisted of layers of lead and rubber. Lead thickness varied from 1/4 inch to 2 1/2 inches, and the rubber thickness ranged from 7 to 17 inches. Only the pilots could see outside, and visibility was provided with a combination of leaded glass and Plexiglas about 10 inches thick. The compartment was pressurized and air conditioned. The main entrance hatch was a hydraulically closed 500- pound door that would fall open when unlatched in an emergency. In addition to the crew shielding compartment, an 8,000 pound lead disk was installed in the center of the fuselage in front of the nuclear reactor. This would block most direct radiation in the direction of the crew cabin. In addition to the lead, the reactor had water shields as added protection. Each of the nine water shields could be drained or filled, depending on the mission.

The massive 35,000- pound nuclear reactor was hung in the aft bomb bay, by a single hook, which could be opened to jettison the reactor in an emergency. What a thought a nuclear reactor plunging to Earth! Look out below! Of course, this never happened. No mid section beefing up was necessary, since the bomb bay was designed to carry

1000KW, 20 Ton Test Reactor, Below
Built by Convair to test reactors in flight

Reactor is installed in bomb bay (top) during "cold" practice insertion. During insertions, all personnel were stationed in shielded control room (note men at rear).

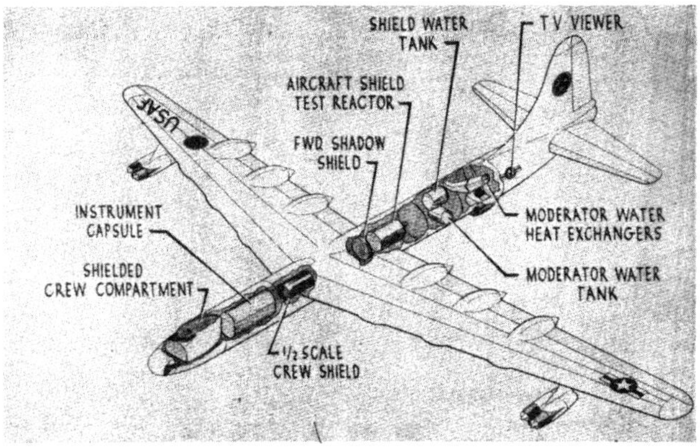

Airborne reactor is located in B-36H fuselage just aft of the wing. Television cameras were installed for engine scanning; note extensive shielding around crew compartment.

Data was recorded in an instrument capsule in the forward bomb bay. The main test data were the operational parameters of the reactor, plus the many radiation monitors throughout the aircraft. 27,000 data points could be captured in a 12 flight.

The reactor itself was designed by Convair. It was flexible enough to operate on the ground and at altitudes of 40,000 feet. It was water-cooled, and its core was made up of fuel rods containing enriched uranium. The reactor had three control rods to moderate the power output. Designed solely as a radiation source for working on shielding and handling problems, the reactor had no output as an actual aircraft propulsion system. Its power level was minimal compared to what would be required to operate as a power plant for an aircraft this size.

Maintenance and operation of the reactor was carefully planned and carried out in an isolated area in West Texas. Facilities and procedures used by Convair would be quite valuable if the program went into production. The reactor was loaded and unloaded for each flight. The reactor was kept in the bottom of a handling tank that could be filled with water to protect personnel. It rested on a rotatable cradle which could be tilted to access each side of the reactor. All work was done by remote handling tools. The roof of the building holding the handling tank could be rolled back, and a gantry crane would pick up the reactor. The gantry crane operator sat in a shielded cab, and moved the reactor to the loading pit near the aircraft. The reactor was installed on a hydraulic lift platform. The nuclear aircraft was towed to the loading pit by a shielded tow vehicle.
One wheel was chocked on a turntable, and the aircraft was towed around until it lined up with the lift platform. Then, all personnel moved to a shielded area during the 20 minute loading operation.

The NB-36 nuclear test aircraft made its first flight with an operating nuclear reactor in September 1955. Several hours elapsed between loading of the reactor and takeoff. Final checkout of the airplane, reactor, and nuclear instruments was made during this period. The NB-36H was taxied from the reactor isolation area, the crew checked for flight readiness. The NB-36H then took off for its historic mission at a maximum take-off weight of 367,000 pounds. The test aircraft had to fly a prescribed course around West Texas and New Mexico. During this time, the reactor engineer prepared the reactor and auxiliary systems for operation, while the flight test engineer prepared the data- gathering gear. Flights were monitored from a Boeing B-50 (improved B-29) chase plane, which could observe the NB-36 in flight. The B-50 also carried nuclear instrumentation for mapping radiation fields around the nuclear reactor aircraft. An Air Force transport also accompanied the aircraft on its test flights, carrying a team of paramedics who could parachute to the ground to isolate and monitor any area where the reactor might have to be jettisoned in an emergency.

When the aircraft arrived over the test area, the core of the reactor was filled with water, and the control rods were withdrawn to start it operating. Data was taken at several configurations. One flight was made outside the normal test corridor. A low level run was made over the Gulf of Mexico. Air scattering data was taken at an altitude near sea level, yet the aircraft was high enough to avoid ground scattering effects. Air scatter data was obtained by the trailing B-50 aircraft. B-50 pilots found they could maintain range between the aircraft by watching a radiation monitoring instrument, which showed changes in radiation activity. When a data run was complete, the reactor was shut down by inserting the control rods. Cooling water was circulated during the return trip to remove some of the heat built up by the reactor's power.

During a program of 47 flights, the aircraft explored the effects of air scatter and radiation patters produced by the aircraft in flight. The last flight was made in March 1957. Convair's successful flight test program showed that the aircraft posed no threat, even if flying low. The principal concerns would be: (a) accidents which cause the release of fission products from the reactors, and (b) the dosage to crew and ground personnel from exposure to leaking radiation.

In 1958, the aircraft was towed away to be stored, after it was apparent that no more USAF money would be supplied to support more NB-36 testing.

In addition to the NB-36 test program, the US Air Force contracted for broad, basic investigations of materials and components in relation to performance in a nuclear flight environment. These included the Nuclear Aircraft Research Facility (NARF), built and operated by Convair at Fort Worth, Texas. An extensive radiation testing program was conducted to find materials and components particularly suited to the new environment of the nuclear aircraft. A major testing device at NARF was the ground-test reactor, capable of dynamically testing small aircraft components in a nuclear radiation field. Shielding experiments were also carried out. Handbooks were written to consolidate the data gathered.

Of course, the GE-ANP Department was also conducting extensive tests, especially at the Arco Test Station, on materials to be used in the reactor.

Another facility was the Tower Shield Facility at the AEC's Oak Ridge Plant. This consisted of four 324-foot towers, with steel-cable assemblies capable of lifting large weights. The purpose of this facility was to simulate reactor radiation patterns during flight. At the same time, numerous mock-ups of different crew shields and shield materials were evaluated, by raising them up in the vicinity of the reactor. Radiation dose rates for different power levels at various heights above ground were recorded.

In Georgia, the USAF contracted Lockheed Aircraft Corp. to construct a radiation effects facility to supplement the NARF. This facility would irradiate large aircraft structures and major subsystems. It was capable of irradiating items as large as 10 by 15 by 8 feet. In addition, the USAF built a nuclear radiation facility at Wright-Patterson AFB, in Ohio. This facility contained a 10-megawatt research reactor and two remotely operated test cells, for further nuclear effects research.

1950's photo of NB-36H test reactor being lowered into rampit by crane. The ramp pit shielded the flight test reactor between flights.

G E Nuclear Reactor

First, a primer on the "direct cycle" concept. This is the approach that General Electric took, and major contracts were provided by the US Government to develop this concept. In this approach, air enters the jet engine and is compressed by a standard compressor. Behind the compressor is a valve which channels the compressed air into the nuclear reactor core. This air is rapidly heated by the nuclear heat from a controlled chain reaction. Then, the highly pressurized hot gas is channeled back into the jet engine to drive the turbine, and also produce thrust. The reactor has simply replaced the chemical burning in the standard combustion system. The air that passes through the reactor core also "cools" off the heat from the nuclear reactor. This is in addition to the moderator in the reactor (usually water) to cool the reactor.

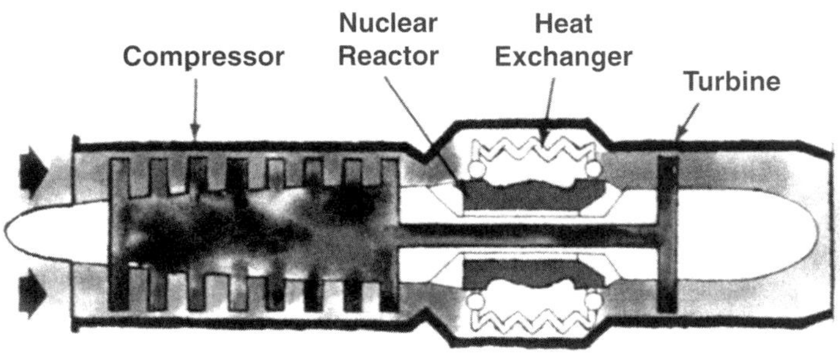

Direct Cycle Nuclear Turbojet Engine

The ducts to flow the air into the reactor core were composed of thick castings. The "in-direct cycle" system, pursued by Pratt & Whitney, was similar. However, air does not pass through the reactor itself. It passes through a heat exchanger. The heat exchanger contains water or other fluids that circulate through the reactor core. In any event, the air is rapidly heated as it passes through the heat exchanger. Same result - thrust generated by the turbojet from the nuclear reaction.

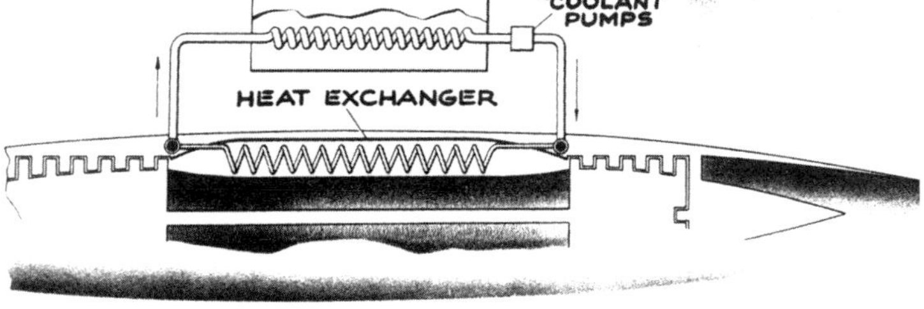

Indirect Cycle Nuclear Turbojet Engine

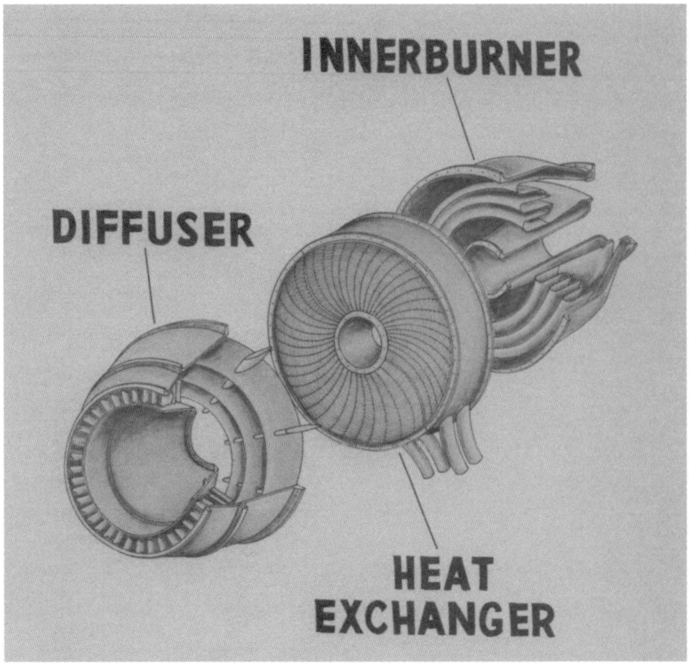

Above is a General Electric proposal engine (X-310) circa 1959, for an indirect cycle engine using a circular heat exchanger. Note the Triple Annular Combustor

GEAE Drawing

The characteristics peculiar to aircraft nuclear reactors might be listed as follows: high power, small size, low weight, high temperatures, and operational reliability. It is estimated that an operational nuclear powered aircraft would weigh about 500,000 pounds. For such an aircraft to fly at Mach 0.9 at 35,000 feet, the reactor power is calculated to be in the neighborhood of 300 megawatts.

Reactor size is an important consideration for aircraft use. The reactor, including shielding, must be of a size that can be incorporated within a streamlined airframe. Reactor size is in large measure the governing factor in nuclear aircraft weight. The larger the reactor, the larger and heavier the shield must be that protects the crew from radiation. The tremendous weight of reactor shielding is one of the main considerations in the successful development of the nuclear aircraft. An aircraft reactor must be a high temperature reactor. For a given size and weight, the higher the temperature, the greater will be the power output. In the conventional jet engine design, we have continually striven for the highest possible combustion chamber exhaust temperatures to produce the highest efficiency. So too in nuclear reactors, we must strive for the highest output for a given size. This is why material properties and design criteria are so important.

At the beginning of the program, there were no materials available that would: (1) stand up to the high-intensity nuclear radiation which necessarily existed throughout the interior of the nuclear reactor, (2) resist corrosion by the very hot air which passed through the core of the reactor at great speed, (3) be guaranteed not to leak any of the highly radioactive fission products into the exhaust air-stream.

The Convair X-6 and the GE XJ53

In early 1951, plans were in place to modify a B-36 bomber into a proof of concept aircraft. It should be noted that at no time was the X-6, as the Modified B-36 was to be called, considered a contender for an operational configuration. No tactical requirements were imposed on the X-6 during the design phase. It was optimized for test bed use only. The original engine concept was to use a liquid indirect reactor to engine lashup. The proposed engine was an engine called the GE XJ53 Turbojet that produced 17,950 lbs. thrust.

In the late 1940s and early 1950s, with the J47 well into development and production, Aircraft Gas Turbine division management ordered a study of future military and civil aircraft planning and the powerplants required. Four completely different approaches evolved from this study.

In 1947 the Air Force had funded Pratt & Whitney for an engine study that eventually resulted in the J57 turbojet. At the same time the USAF planners and Power Plant Lab also began to look for a very high thrust engine. From this interest came GE's XJ53, a 1948-originated design calling for a thrust output of nearly 17,000 pounds-double the thrust of most engines in existence at that time.

Projected weight of the engine was 6,000 pounds-3,500 pounds more than the J47. Both the thrust and engine weight made it by far the most powerful and largest jet engine of its time. It was, of course, an axial flow design.

When the first design was completed and projected weight indicated a total of more than 8,300 pounds, the engineers knew they had a problem. By the time the first engine ran on test in March, 1951, a combination of Lynn and Schenectady engineering experts had succeeded in reducing the weight to 7,950 pounds with the engine delivering the unprecedented power of 17,950 pounds of thrust. However, it was painfully apparent to both GE and the USAF that the engine was entirely too large-no airframe then contemplated called for an engine of that size-and that further efforts to reduce weight would result in unsatisfactory performance characteristics. The XJ53 engine development was discontinued in September, 1953.

By June of 1951 the laidout plans called for a test stand powerplant by 1954 and a nuclear flying test bed (The X-6) by 1956 under an AEC contract dated April 30, 1951. Later in 1951 GE proposed that its powerplant design be changed from the proposed liquid metal cooling cycle reactor design to a direct air cycle system. Formal approval was endorsed by the Airforce during October 1951.

General Electric assumed throughout the summer of 1951 that a direct cycle, aircooled reactor would require fuel elements operating at some 2,500°F in order to produce a respectable amount of thrust. Since no known materials could long sustain such temperatures, a design change to 1,800°F for fuel elements was suggested (even though take off would have to be on conventional turbojet power) for the first nuclear powerplant. The lash up of Reactor R-1 and the Turbojets (4) became the P-1 Powerplant.

R-1 Powerplant Reactor

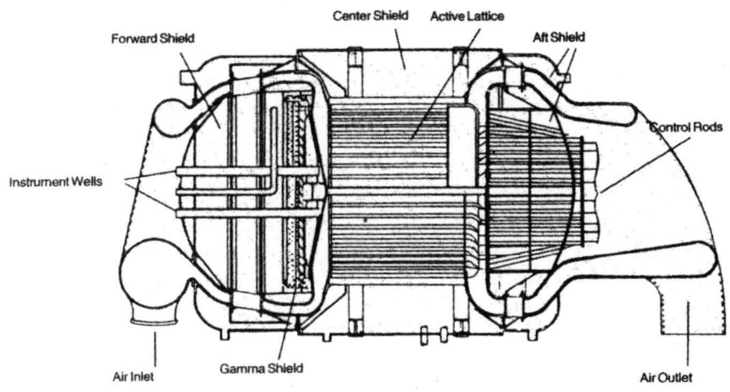

P-1 Powerplant
with reactor on top

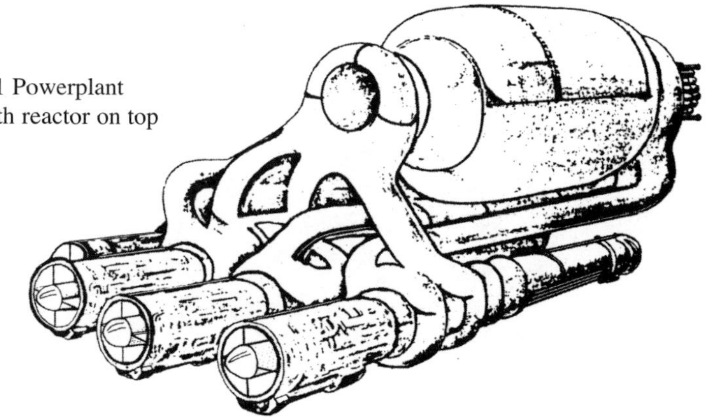

The plan was to use 4 GE modified Turbojets (J47 or XJ53)

Additional studies soon indicated that existing General Electric J47 engines (4) connected to a direct cycle aircooled water-moderated reactor with 1800°F fuel elements would most likely propel the X-6. The four J47 (now designated X39) would have to produce a total of 26,000 lb Th at 15,000 ft. using reactor heat alone, but would have additional combusters for chemical fuel for take off use. Max speed was to be 300 to 390 mph for the X-6.

The design for the reactor was to deliver 150 megawatts of power. The design called for the reactor to be contained within the aircraft, with the propulsion engines slung below. Total weight for the system would be:

reactor core	10,000 lbs.
four engines	18,000 lbs.
air ducting	40,000 lbs.
shielding	60,000 lbs.
crew shielding	37,000 lbs.

Total weight was 165,000 pounds. The reactor core was 68 inches in diameter. It was to be manufactured of stainless steel, with uranium oxide fuel. The fuel cartridges were constructed in rings, with the compressor air flowing between the rings. The shielding was 10 feet in diameter and 20 feet long.

P1 Contender Engine, The GE XJ53 Turbojet produced 17,951 lbs of thrust in 1951 at a weight of 7950 lbs.
GEAE Photo

P-1 and R-1 mockup for the Convair X-6. This assembly would fit into the belly with the reactor inside the aircraft and the engines slung underneath.

Below is a model of the Convair X-6
Note: The four nuclear powered jets slung under the belly. The R1 reactor was inside the bomb bay.

For additional thrust GE was in a study phase to use PW J57 to replace the GE J39 to be called the X40.

Unfortunately, the P1 program was canceled by the incoming Eisenhower administration in 1953. When the NB-36H did fly in 1955, with a non- GE nuclear reactor on board, there were no jet engines installed. The only information gathered concerned the effects of radiation and aircraft/reactor handling. The eleven X39 engines were later used at the Idaho Test Station for testing with the follow-on HTRE series of GE-produced reactors.

X-39 (J47) Tested with Remote Burners

The Program Moves Forward

The HTRE series of GE- designed reactors

HTRE now stands for Heat Transfer Reactor Experiment. There were three configurations, each more advanced than the previous one. Work began on HTRE-1 in September 1953. It consisted of a nuclear reactor, a radiation shield, two jet engines on twin railroad tracks, ducting, control components, a chemical (jet fuel) combustion system, accessories, an after-heat removal system, and necessary instrumentation to capture the test results. A mock-up was completed in August1954, followed by drawing release for manufacturing in September, and manufacturing was complete by August, 1955. These reactors were all built up and tested at the Idaho Test Facility. In November 1955, the HTRE-1 first went critical: that is, produced power from its nuclear reactor. Full power was achieved in January 1956, and initial tests were completed by January 1957. Contracts signed in 1951 called for jet power using a nuclear reactor by January, 1956. On January 31, 1956, GE succeeded - the HTRE-1 reactor powered the twin X39 jet engines.

HTRE-1 (circa 1955) Full power achieved Jan 31, 1956 at Idaho test site.

INEEL Photo

The engines were always started on chemical fuel. As the reactor heat increased, the jet fuel was decreased to maintain a constant temperature in the jet's turbine. Finally, when the reactor's heat was sufficient, the jet fuel was stopped. The reactor operated for 5004 hours, producing up to 20.2 megawatts of power. Full -power runs totaled 151 hours; the goal was for 100 hours at full power. The output of the reactor was channeled through two J47 engines, modified to the X39 configuration. This was the first known jet engine operated by nuclear power.

Thirty-seven fuel cartridges of uranium U-235 pro duced the power using a nicrome clad fuel element of concentric ring design. There were also control rods to moderate or stop the reaction. Water to a maximum of 160 F moderated the nuclear reaction. Moderators are used to slow down the nuclear reaction. Water, beryllium, zirconium, and carbon were typical moderators. The operating temperature in the reactor was 1700°F. No attempt was made to restrict the size and weight of this equipment to approximate a flight test version. Rather, the assembly was made large for ease of access.

The rebuilt HTRE-1 reactor was called the HTRE-2 reactor design. Work began on it in early 1956. Testing commenced in December, 1956, and the reactor operated until the 1961 program cancellation. It was also water- moderated, with 30 fuel cartridges. Same fuel uranium U235. A thicker reflector was added for insulation. These reactor cores were relatively small - 30 inches in diameter, with fuel cartridges that were 30 inches long. The reactor core contained, a hexagonal center section, which could accept modified fuel configurations . The objective was to produce maximum power, while developing fuel cartridges configurations which would survive over many hours of testing (or later military flights). Nine different configurations of fuel rod cartridges were developed. These had varied materials (zirconium, ceramic, beryllium, steel); some had coatings applied to the inner and outer surfaces; different shapes and configurations were tried out. The results were not always encouraging. Some configurations developed meltdown, or fuel tube burning, or tube blistering, or just not enough power. Finally, a configuration (cartridge design L2E6) worked. It operated for over 300 hours at temps approaching 2000°F. The fuel elements looked good after the tests.

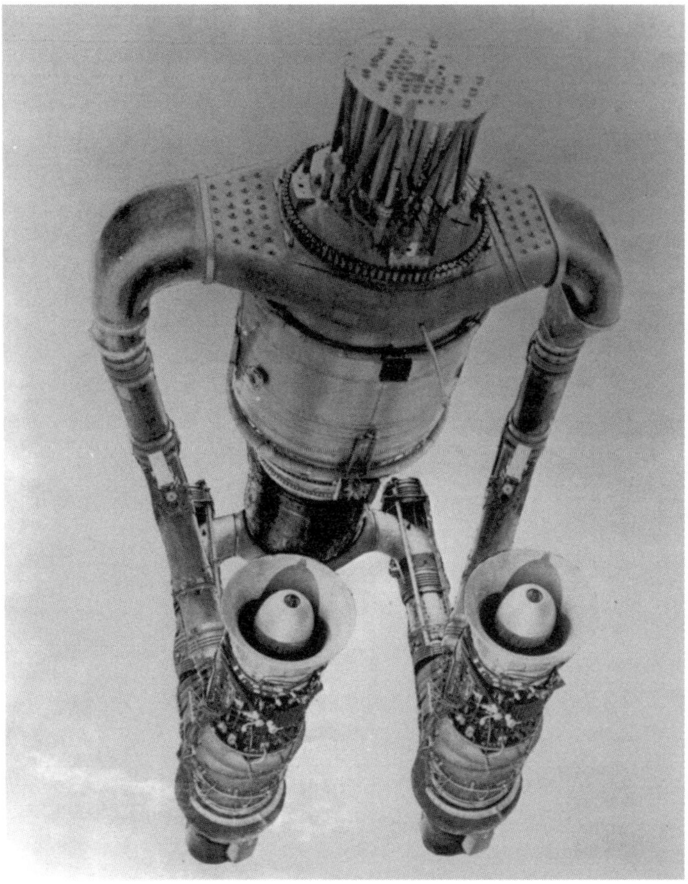

Above is the HTRE-3 "Gut" Engine less shielding, etc. Note: The X39 (J47) Twin lash up. This is much like the P 3 engine system envisioned for the X-6 A/C. Except it called for 4 engines for the Corvair X-6 A/C.
GEAE Photo

Above: The General Electric Evandale Ohio Plant. Built by Curtis Wright A/C Company during WWII. Photo Circa 1955. It was at this plant that all nuclear engines were designed and tested with chemical fuels only. All "Hot" testing was done at the Idaho Test Site.
GEAE Photo

The follow-on HTRE-3 reactor was to evaluate and confirm performance tests - this was to be a production configuration design. Two X39 jet engines were operated with the reactor. The engine set-up was so large that it was pushed to the reactor site on a duel set of railroad tracks. GE had to develop methods to remove all of the engine components by remote control. This was after they were “hot” having run near the nuclear reactor. This reactor contained hydrided zirconium as an inside moderator (prior reactors used water). A new- design inlet plenum was to carefully meter air from the jet engine through the reactor, and back again to the engine. The core configuration was the concentric ring design using nicrome clad fuel sheets fabricated into tube like cartridges and reflected the best of the HTRE-2 designs, but was modified to improve performance. The reactor was 51 inches in diameter, 34.7 inches in core length and 43.5 inches in over all length with the beryllium reflector, all made up of 151 hex-shaped moderator cells of unclad hydrided zirconium with 3 inch bores for the fuel elements.The fuel elements cartridges were of the concentric ring design,19 stages each 1.5 inches long .The fuel element sheet rings used 80Ni-20Cr alloy cladding over 93% fully enriched uranium cores.(this is over 15 times the enrichment used in a nuclear powered generating reactor) The reactor control system consisted of: control rods to moderate power; a temperature control system to moderate the airflow, and a "scram" mode for immediate reactor shutdown during emergencies.

Testing of this final HTRE series was begun in April 1958. These were low power tests to prove the design concept. By November 1958, overall power plant testing began. Shortly thereafter, an over-temperature condition occurred which melted some of the Uranium fuel rods. It turns out that erroneous test equipment and not the reactor design, was the cause. However, it took six months to repair the reactor core. In June 1959, testing resumed. By October, a power level of 10 megawatts was achieved, with no problems being encountered. By December 1959 , a 166 hour endurance run was accomplished, followed by a full power run of 30 megawatts. A subsequent run of 66 additional hours at full power was accomplished. The program was quite successful. Much new operating data were obtained. Tests continued into 1960.

The overall objective had been achieved. This reactor / jet engine configuration would work for nuclear -powered aircraft flights. The engineering team was filled with pride and satisfaction in achieving the design goals set years earlier. This concept proved that nuclear- powered flight was possible at last!

Photo below shows HTRE-2 (left) and HTRE-3 (right) as they sit at Arco Idaho.
Note the size difference and also the two J39 (J47) engines protruding to the left on HTRE-2

DEVELOPMENT OF THE X211 JET ENGINE

GE's Evendale Ohio plant produced most of the components for the jet engines and the nuclear reactors. Reactor components were flown to Idaho for assembly. The Evendale shops included: standard machine shops, a graphite shop, a special metals shop (with filtering for removal of hazardous materials), a high bay (for assembly), and a clean room (with workers wearing nylon outer garments). Machines included: Electron beam welding, electrical discharge machining, TIG welding (with welds X-rayed), precision jig borers, temperature- controlled inspection areas, and a 12 foot by 12 foot by 20 foot high temperature furnace.

Up to that time, all calculations were made using office calculators - the hand cranked variety. It would take hours to produce a calculation that can be done today in nanoseconds. A new tool, during the late 1950's, for compressor testing was the "ICPAC" program, using an analog computer, which instantaneously depicted compressor performance trends.

The General Electric jet engine developed for the Convair NX-2 bomber prototype was the X211 (military designation was the J87). German- born Bruno Bruckmann was manager of the engine program. Start-up of the design phase was in June, 1955.

The GE task force looked at two configurations. An "in-line" version consisting of a single- engine arrangement with the engine inline with the nuclear reactor. Also, a twin version consisting of a two - engine arrangement. with the reactor between the two engines. In this version, two large ducts or scrolls connected each of the two engines to the reactor, one conveying the compressor discharge air into the reactor, and the other returning the hot gases to the engine. Both versions had an annular combustion system. The other major components of the system were similar to a conventional jet engine. The selection was for the twin engine design, with the nuclear reactor fit between the two jet engines.
Manpower - A total of 33, including 18 engineering people worked in this project at the end of 1955. By the end of April 1956, the personal had increased to 52 people. By the end of 1956, the total was 89. By January of 1958, the number had increased to 120 people.

In early 1956, the configuration was finalized and layout studies were completed. During early 1956, comprehensive design reviews were conducted for the purpose of identifying and reducing design complexity and risk. In late 1956, drawings were released for the first full scale mock-up. In early 1957, the detailed mechanical design for the first block of factory test engines was completed, and all parts for this block were ordered. Early in the third quarter of 1957, the first individual component tests were run on actual engine parts. Initial component testing was begun in October, with the compressor rotor. Considerable engineering effort was expended in 1957 in manufacturing and assembly preparations. As parts were received and assembled however, several design and/or manufacturing deficiencies became apparent.

X211 Twin Nuclear Turbojet

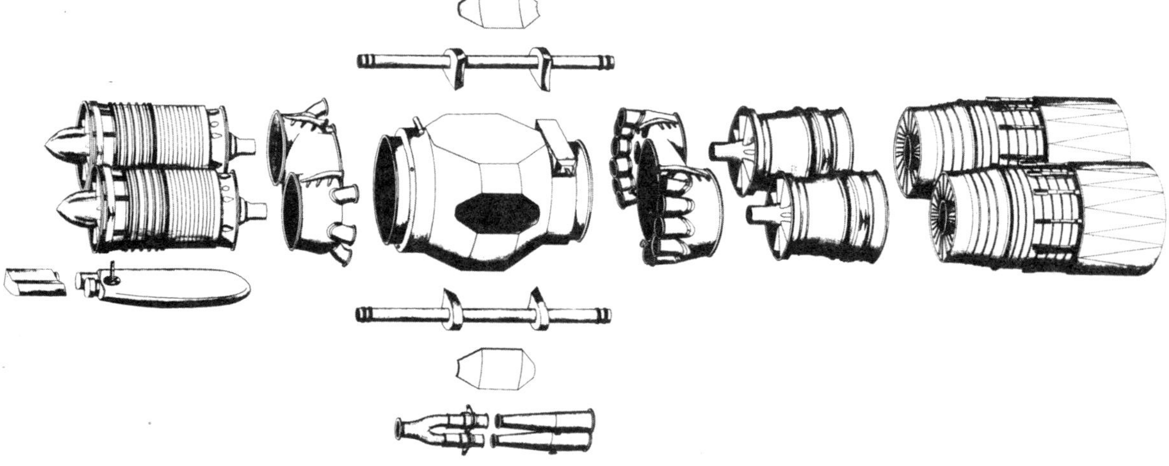

Key sub assemblies of the GE-X211 41 FT. long - 54,000 lbs of Thrust

Above: Is the GE-X211 Nuclear engine in build up phase. Note missing external combustors. See sketch to left. In this buildup only one compressor was working. The unit to the right was a dummy.

GEAE Photo

The general configuration of the X211 engine was established, differing from conventional turbojet practice through the use of exhaust collectors, a by-pass combustion system, and an external compressor-turbine coupling shaft. There were 16compressor stages, with variable compressor vanes. The first design had six variable stages, later engines had all 16 stages variable. Three gearboxes: inlet, transfer, and rear. The turbine rotor contained three stages, with two structural frames, connected by a 17foot long turbine shaft. The exhaust section contained a variable exhaust nozzle (convergent/divergent). In front of the compressor was an air turbine starter. There was a compressor exhaust collector, a unique feature. This supported the compressor and ducted the airflow to either the chemical (jet fuel) combustion system or to the heat exchanger (nuclear reactor). The design also allowed use with the nuclear reactor. In this instance, there were to be just 12 jet fuel burners (on the outside of the engine - away from the reactor). This was for takeoff power and other chemical power during flight. Nuclear start-up was to take place during cruise. The inboard burner cans were replaced by the nuclear reactor, which supplied power in flight. The valve between the compressor and combustion sections diverted airflow to the six jet fuel burners and/or the reactor. Power would be diverted from the jet fuel to the reactor as it powered up. There were actuators on top and below the engine to divert the airflow.

Engine Statistics

Airflow	425 pounds/sec
Compression Ratio	14:1
Shaft Speed (max..)	5,000 rpm
No. of main bearings	six
Weight (guarantee)	15,745 pounds
Length	510 inches
Inlet diameter	55 inches
Nozzle discharge diameter	80 inches max.
Compressor discharge temp.	1200 degrees F
Compressor discharge press.	245 psi
Fuel type	JP4
Turbine inlet temp.	1800 degrees F
Design life	1000 hours

New problems were associated with the X211. The dominating design considerations established for this nuclear engine, in order of important, were as follows:

1. reliability
2. thermodynamic performance
3. frontal area
4. weight

Engine size was also a problem. The X211 engine would be the largest known aircraft gas turbine in the 1960's world. It would be almost twice the diameter, and two times the length of its predecessors. This great size, determined by the heat exchanger (reactor) requirements, has magnified the usual turbojet design requirements. Many new material developments and applications were required in the solution of these problems.

First Engine To Test (FETT) for this concept was at the Evendale plant in January of 1958. This consisted of one engine only. (The twin configuration was tested later in January 1960). Airflow was 425 pounds per second. Maximum power was 27,370 pounds thrust using jet fuel - a GE record to that point. The turbine inlet temperature was 1,700 F. The length of the engine was 42 feet (huge), and maximum compressor diameter was 54 ° inches. Key structures consisted of: a front frame, compressor casing, rear compressor frame, a mixer valve, combustion section, turbine front frame, turbine casing, and exhaust nozzle assembly. Maximum weight was 15,745 pounds. The number of engine builds tested was six. A new test cell, X-1, was built in Evendale in Building D, for the X211 engine. This was the world's largest test cell at the time. Only jet fuel tests were done in Evendale. The Idaho Test Station was the site of nuclear reactor powered tests. Many component tests were run between 1955 and 1957, to test components - the compressor, the combustion system, turbine, and exhaust nozzle. Tests were performed for compressor and variable vane icing.

Below is a mode of operation diagram for the X211. Note: Gates for diverting the compressor all into the outside combustors or into the reactor.

X211 1/10th scale mockup.
Note: Control rods between the two compressors.
GEAE Photo

X211 Modes of Operation

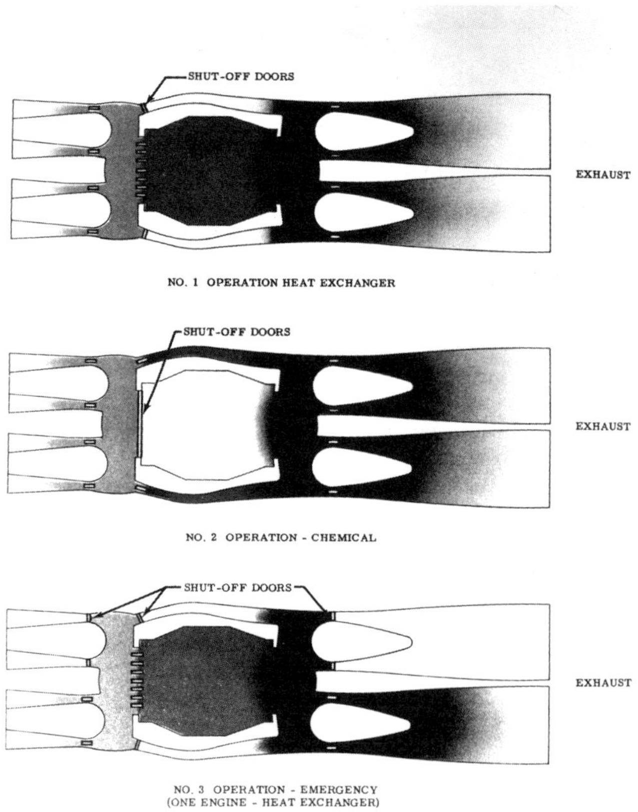

X211 Exhaust Duct for Testing

GEAE Photo

X211 Compressor Testing

Testing was performed at GE's mammoth Riverworks Plant in Lynn Ma. on the X211 compressor during 1957-58. Obviously, problems were uncovered and resolved as the program progressed. Here are some of the problems uncovered and solutions devised.

Build-up One occurred in 1957. Severe mechanical difficulties were encountered during the test program. Failure of the 16th stage air seal resulted in the eventual termination of the testing. However, this first program resulted in the determination of starting vane schedules and aerodynamic performance up to 80% of design speed. A total of 20 variable stator schedules were investigated.

Build-up Two: The 16th stage air seal was redesigned. In the mechanical check-out, an interference noise was heard, and recorded on the second stage rotor strain gauges. This was determined to be caused by a slip ring spline fit. However, tests were ended by a stall in which several sets of blades and vanes were damaged. During this series of tests, 12 vane schedules were investigated, and design speed approached 90%. Although weight flow and compressor efficiency approximated design planning, the pressure ratio was low. Results did allow for further refinement of the design.

Build-up Three: Testing was spent in trying to determine the optimum stator vane schedules up to 100% speed. High stresses occurred, however. This was alleviated somewhat by the detection of a disconnected 9th stage stator vane. This test phase was also ended by a major stall which produced high stresses. However, research data was obtained between 40% and 100% speed. Air flow was as expected, but efficiencies were still lower than planned. It was possible, however, to attain 100% speed with the most advanced compressor yet tested. This allowed the operation of the full jet engine to proceed.

Build-up Four: After initial test runs, the vane angle actuation system was balky. Otherwise, mechanical operations were completely successful. Thirty-one vane schedules were investigated during this phase. Design pressure ratio and efficiency were somewhat lower than expected, while air flow was OK. Overall, the X211 compressor set new records with multiple variable vanes, and a pressure ratio exceeding 20 to one.

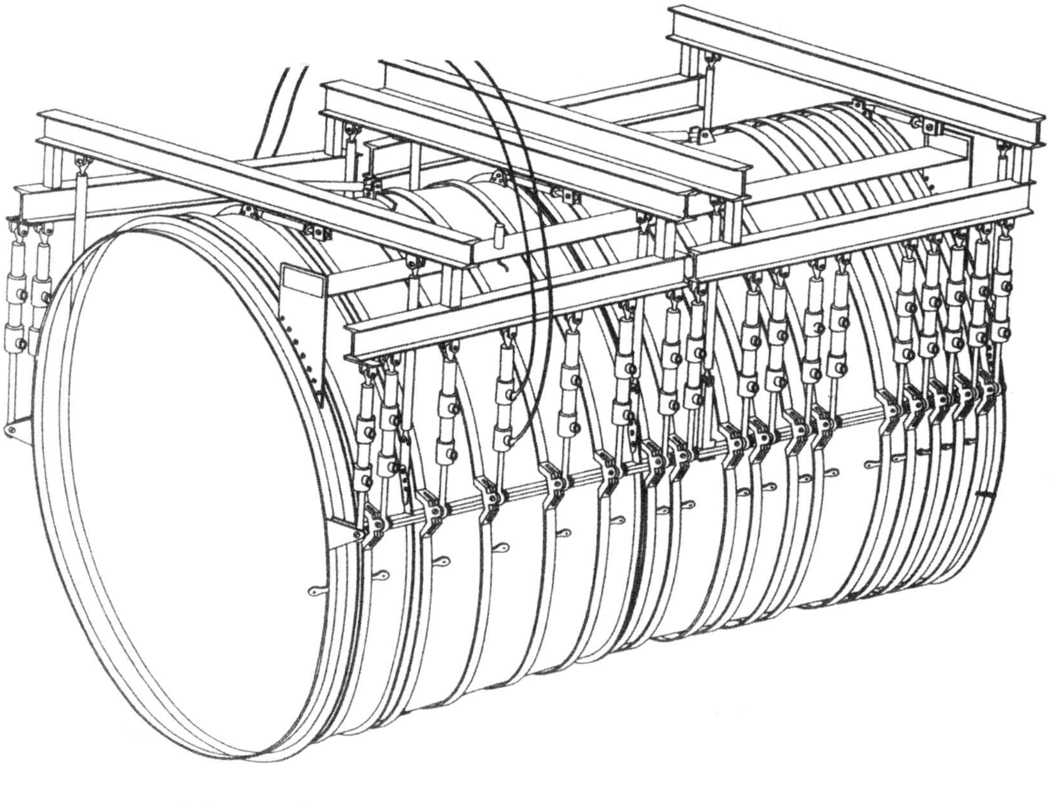

16 Stages of Vanes shown above in this Vane test Rig Sketch.
The X211 set a record for multiple variable vanes, and a pressure ratio exceeding 20 to 1.

To give a sense of the size, this photo shows the exit from the reactor on the X211. Note: The men below. This engine had 42,000 engine parts (not including piping and wiring), as compared to a J79 which had 12,000.

GEAE Photo

Nuclear Reactor / Jet Engine Combination

The XMA-1A was the designation for the reactor and the twin GE X211 jet engines. The reactor was sandwiched closely between the two jet engines. It consisted of a core section (where the nuclear reaction took place). The prototype core was 38 inches long, with a diameter of 62 inches. It weighed 11,900 pounds. Output power was 192 megawatts. It was wrapped around with a side shield (of lead and water), a front plug, and a rear plug (also lead and water). The core consisted of 151 zirconium hydrided fuel elements, using enriched uranium-235 as a nuclear heat source. There were also 129 control rods - seven dynamic rods and 122 shim rods.

The air from the compressor would pass through the reactor to absorb heat. It also cooled the reactor in the process. Then, it was channeled back through the engine turbine to produce compressor power. The residual power from this prototype reactor engine combination produced almost 55,000 pounds thrust (from the two engines). Two of these XMA-1A reactor jet engine combinations were to power the Convair NX-2 bomber prototype, a 450,000- pound aircraft. Speed was "high sub-sonic," The contract called for 1000 hours of powered flight - this is equivalent to a distance of 500,000 miles! It was supposed to fly in 1963, with nuclear -powered flight by 1965.

Convair NX-2 (About the size of a B-52)
450,000 lbs.
(Estimated Weight)

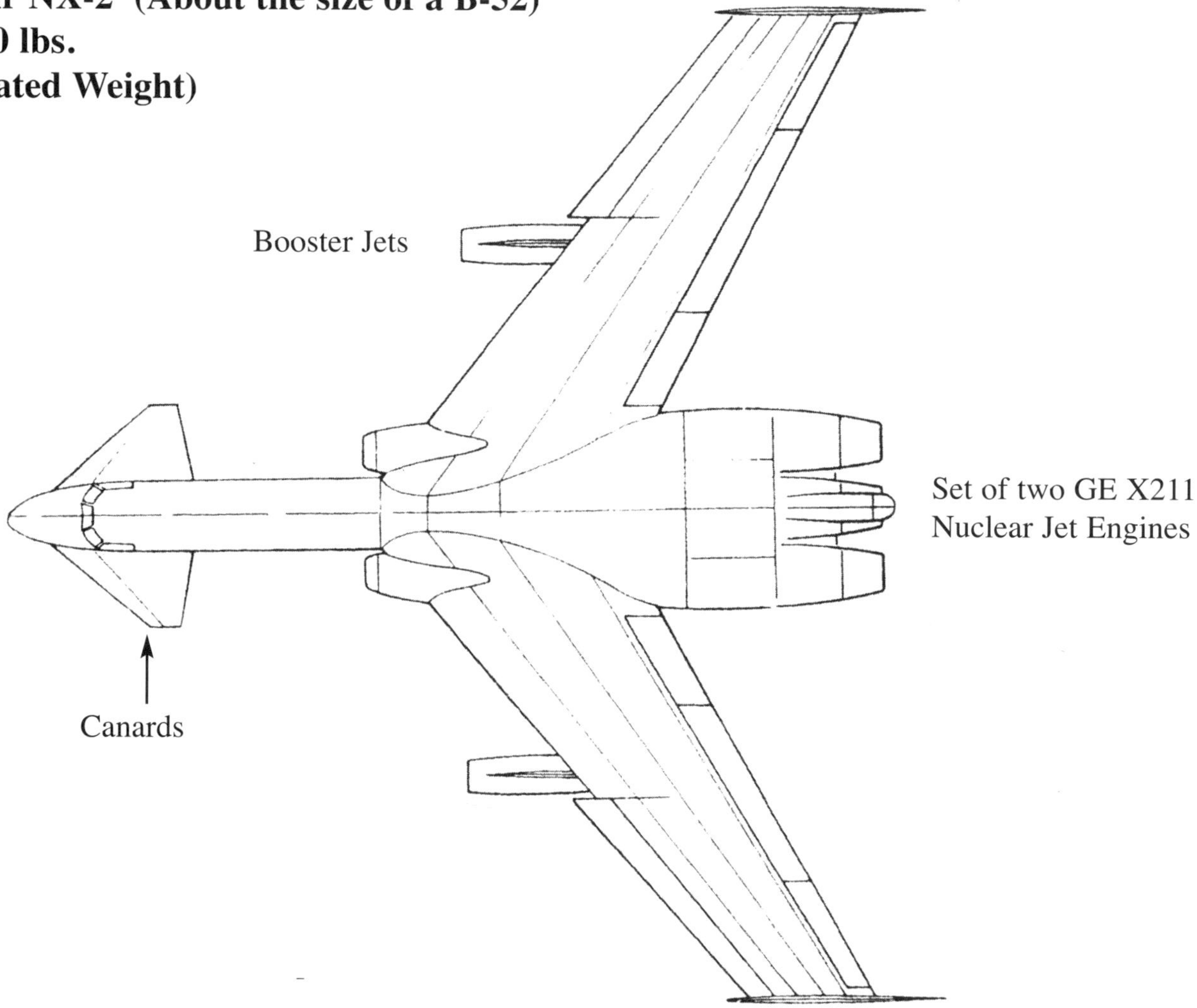

Proposed Convair NX-2 powered by the GE X211 Nuclear Twin pack in the tail with an additional set of two booster jets on the wings.

Additional engines of the X211 series were in planning stages thru the mid to late 1950s. One such engine was the X310. Seen in cross-section below.

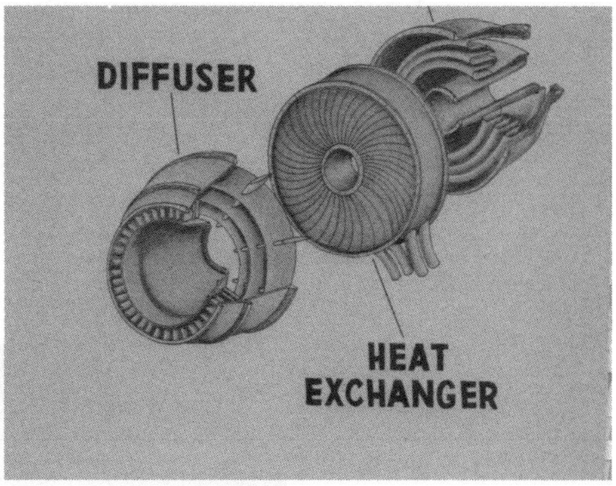

Detail of indirect nuclear heat exchanger for the X310.

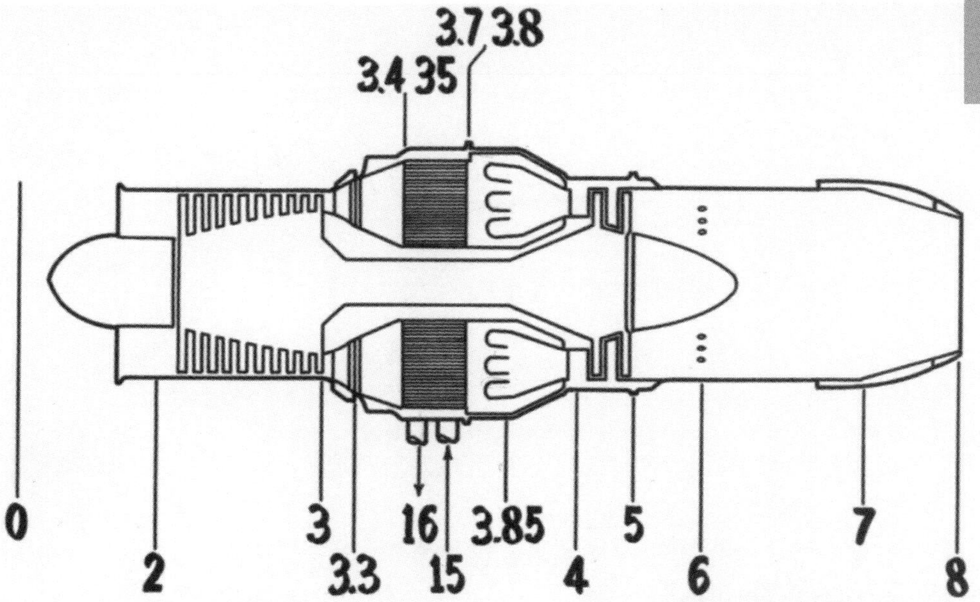

0 FREE STREAM
1 ANY POINT OF PARTICULAR INTEREST FORWARD OF COMPRESSOR
2 COMPRESSOR INLET
3 COMPRESSOR OUTLET
3.3 DIFFUSER SPLITTER SECTION INLET
3.4 DIFFUSER OUTLET
3.5 HEAT EXCHANGER INLET
3.7 HEAT EXCHANGER OUTLET
3.8 INNERBURNER INLET
3.85 POINT AT INNERBURNER DOMES
4 TURBINE INLET
5 TURBINE OUTLET
5.1 POINT WHERE TURBINE COOLING AIR HAS REENTERED MAIN STREAM
6 TAILPIPE INLET (UPSTREAM OF AFTERBURNER FLAMEHOLDERS
7 TAILPIPE OUTLET (JET NOZZLE INLET)
8 JET NOZZLE OUTLET
15 HEAT EXCHANGER FLUID INLET
16 HEAT EXCHANGER FLUID OUTLET

FIG. 2

The X310 planned to use only one compressor taken from the X211 program. It can be assumed its output thrust would have been about 20,000 lbs.

XNJ140: **A single compressor version of the X211. This engine would have been the most likely flight engine if the the program had not been cancelled in 1961. The NX-2 would have used three XNJ140's.**

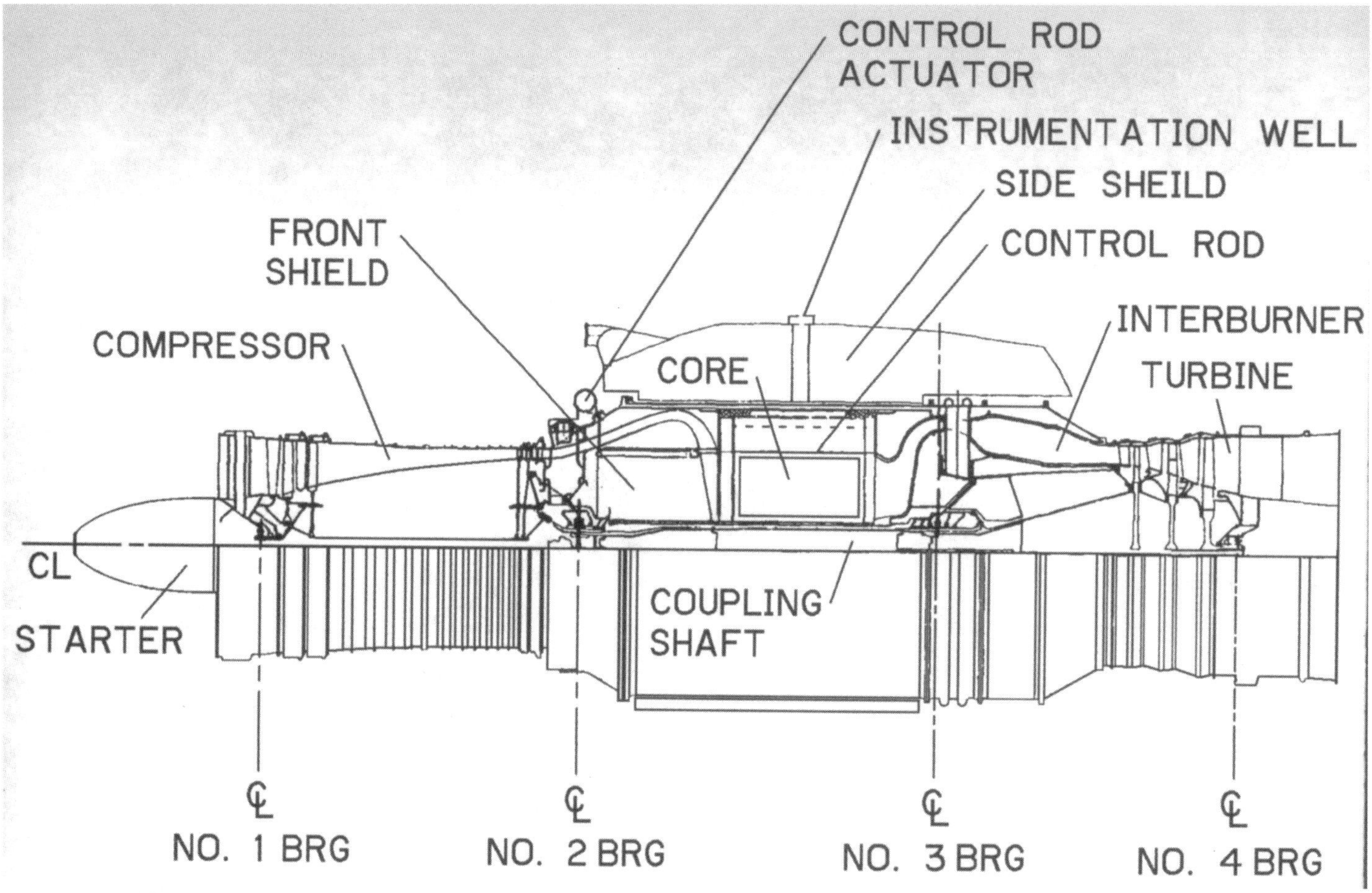

Right: The XNJ140 in full build at the General Electric Plant in Evendale Ohio.
(See page 17 for a comparison to the dual compressor X211.)
Note that the reactor is in direct line with the compressor and turbine. The XNJ140 used the same reactor as the X211.

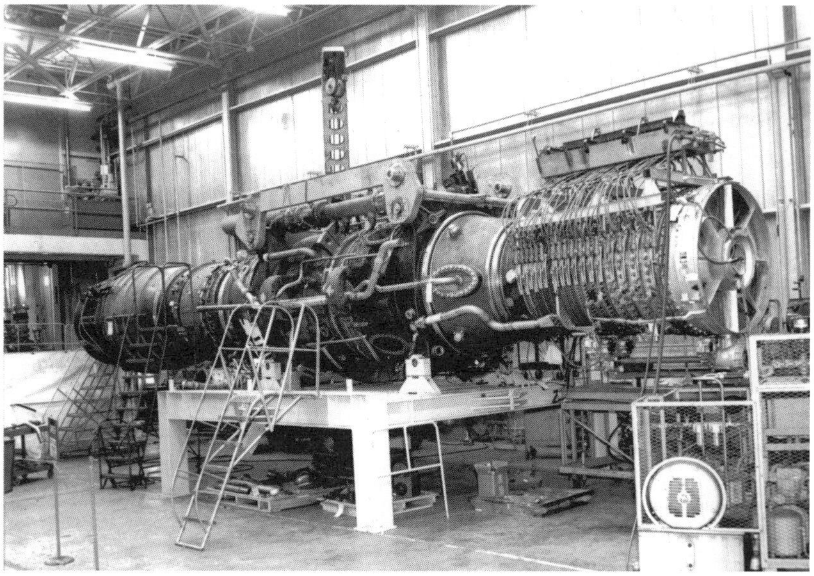

GEAE Photo

Side View NX-2

Top View

Bottom View

Rear View NX-2

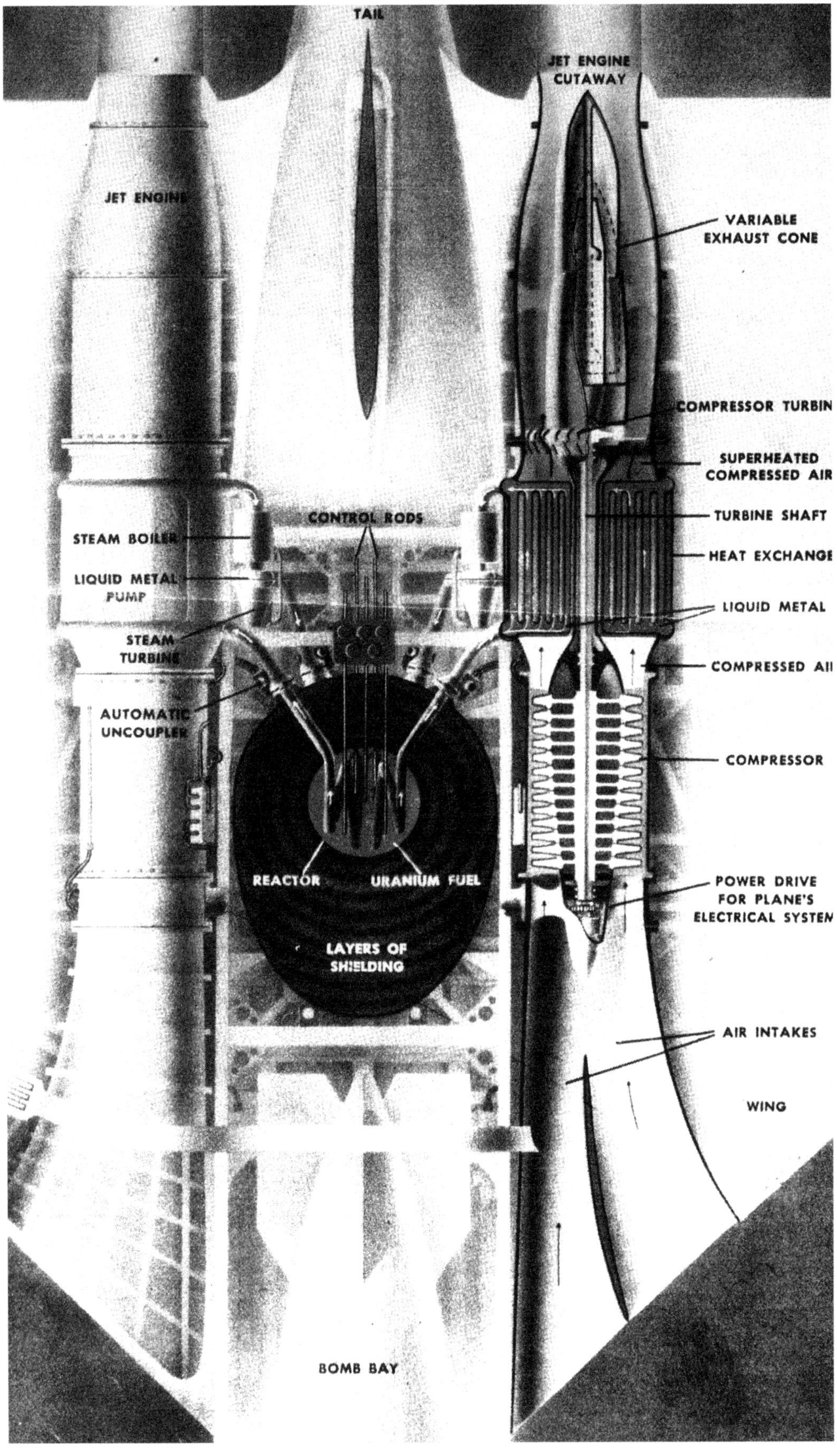

Life ran this drawing of a proposed nuclear engine using the indirect cycle in late 1958. The article went on to describe the workings of the proposed engine.

Atomic Power Plant consists of reactor between twin jet engines at roots of plane's wings. Air (dark arrows) is scooped from outside through vents. it is first compressed, the superheated in heat exchanger where it expands violently, blasting from exhaust ducts and driving plane forward. Exchanger is kept hot by flow of liquid metal (white arrows), heated to over 1,200°F. in reactor and circulated by a steam-driven pump which uses metal's heat to make steam. Turbine, placed in path of hot air stream and turned by it, operates compressor and plane's electric generator. Plane's speed is controlled by control rods which lower temperature when thrust into reactor and by movable exhaust cone which varies size of exhaust opening (dotted line shows its retracted position for top speed) letting more air escape when greater speed is required.

LIFE 1958

Life Magazine depiction of an Indirect cycle nuclear jet engine.

By the late 1950s all sorts of ideas came to the popular press as can be seen below from a Life Magazine article.

AGE OF THE A-PLANE

U.S. MOVES AHEAD ON ATOM-POWERED FLIGHT

Last week General Nathan F. Twining, Air Force Chief of Staff, announced that the Air Force is rushing the development of an atom-driven airplane, a plane which could fly anywhere on earth without refueling. Present work on the A-plane must remain secret. But using only unclassified material, two experts have designed for LIFE a feasible A-plane, shown on these pages. In this drawing three A-planes flash over a landing strip with a B-47 (*directly above*). Another A-plane lands, using a drag parachute. Still another rolls through a slit in a thick concrete wall (*left*) where it disappears into the hillside hangar shown on the following pages.

CONTINUED ON NEXT PAGE

A-PLANE OF THE FUTURE WOULD BE SERVICED

General Twining's historic announcement came only a short while after the atom-powered submarine U.S.S. *Nautilus* came up from its first successful dive (LIFE, Jan. 31). Like the *Nautilus* an atomic airplane could travel enormous distances powered by a lump of uranium no bigger than a man's fist. During the past year the Air Force and the AEC signed contracts with several airplane companies for work on nuclear aircraft. But work on the A-plane has been slower than on the atomic submarine because of the vastly greater problems involved in atom-powered flight.

The greatest challenge confronting nuclear physicists and engineers has been to design a nuclear power reactor which would be light enough to be hurled under its own power through the stratosphere at supersonic speed, yet heavy enough with shielding to protect the crew from radiation. This need determines the plane's wild-goose shape shown on the preceding page. Instead of having its wings toward the front like a B-47, the atomic bomber would have its delta wing at the rear to carry the load of a massive nuclear power plant. In front of the wing would stretch

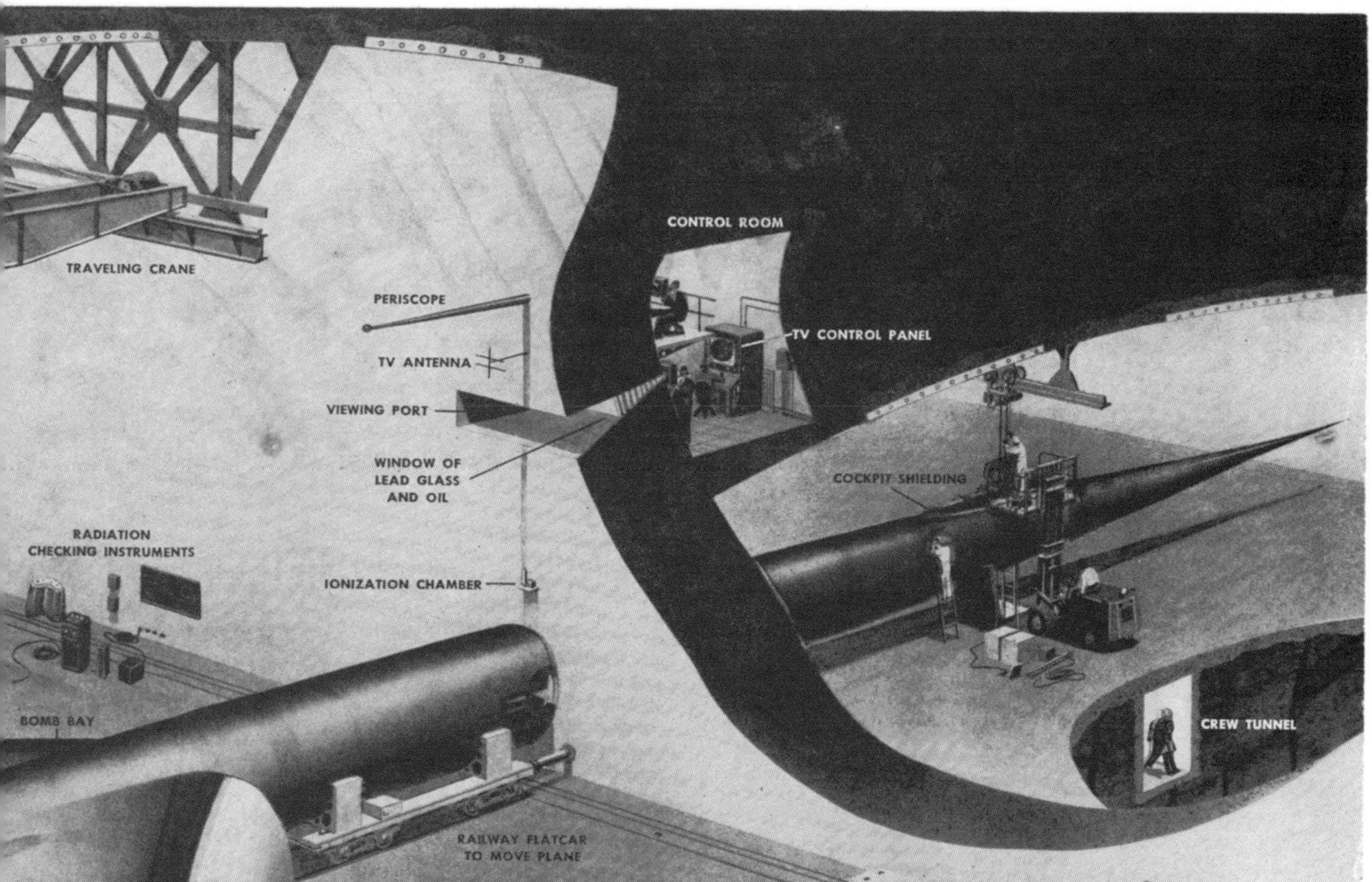

IN A HILLSIDE HANGAR

the hundred-foot-long fuselage whose only purpose is to remove the three-man crew as far as possible from the reactor's radiation.

Danger from the rays would also make it necessary to service the plane in a radiation-proof hangar, in this case a vast chamber (*above*) carved out of a hillside. Here the atom plane's harmless nose would poke into an inner chamber where servicing could be done by hand. The dangerous task of unlocking and replacing the reactor would be carried out in a special "hot" chamber by intricate machinery operated by remote control.

ATOM-PLANE'S HANGAR is an underground chamber in which atomic jet is serviced after its mission. Mounted on a railway flat car, the bomber has been rolled into service area of hangar through slit outer wall (*left*). Once inside, it plugs a hole piercing a second shielding wall (*right*) which protects safe inner area of hangar. In foreground, work on reactor is done by remote control by technicians in control room. They watch through a thick viewing port, a periscope and TV as crane, which has lifted off reactor compartment cover, now lifts out the egg-shaped reactor, which has been automatically uncoupled from jet engines. Ionization chambers measure the reactor's radioactivity. Crane carries the reactor to a pool and immerses it in water which blocks radioactivity. In the inner service area, maintenance men check cockpit instruments. After the reactor has been immersed, crew can safely enter the "hot" chamber to supervise, from behind lead shields, the underwater reloading of the reactor with uranium. After this the reactor is replaced in the plane which is then ready for another mission.

CONTINUED ON NEXT PAGE

ATOMIC ENGINES FOR AIRCRAFT

"More Dreaming"

All the ideas on this page came from a book written in 1952 by C.P. Lent.

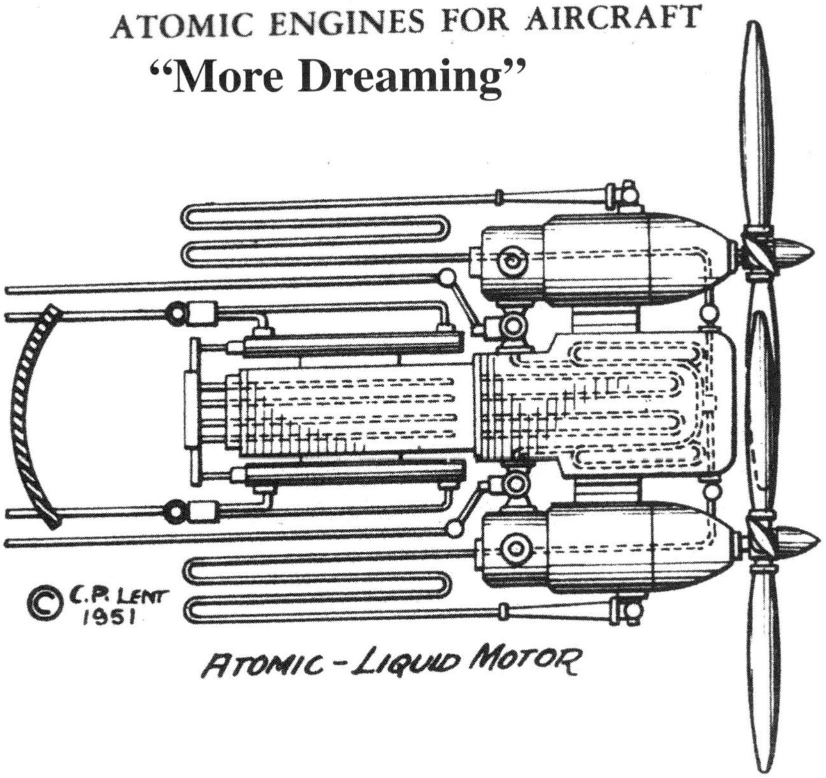

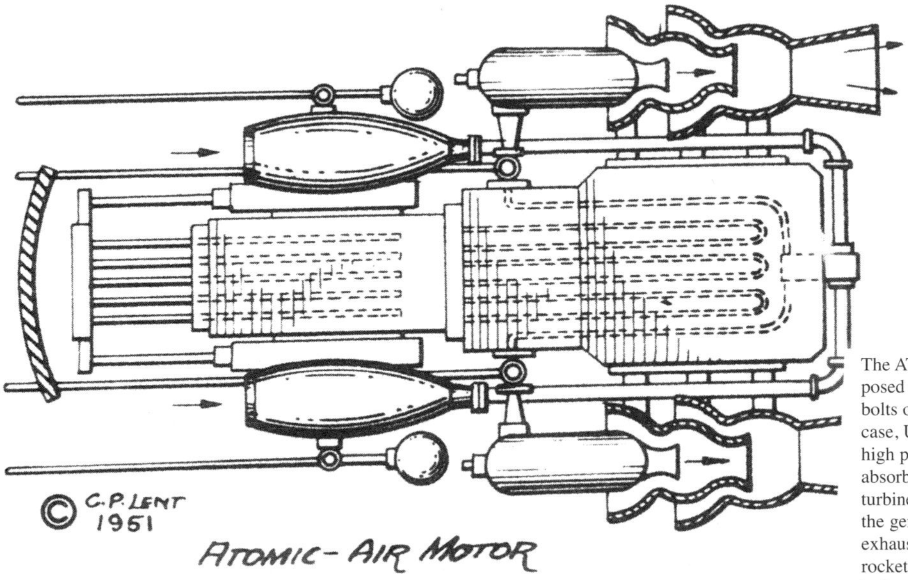

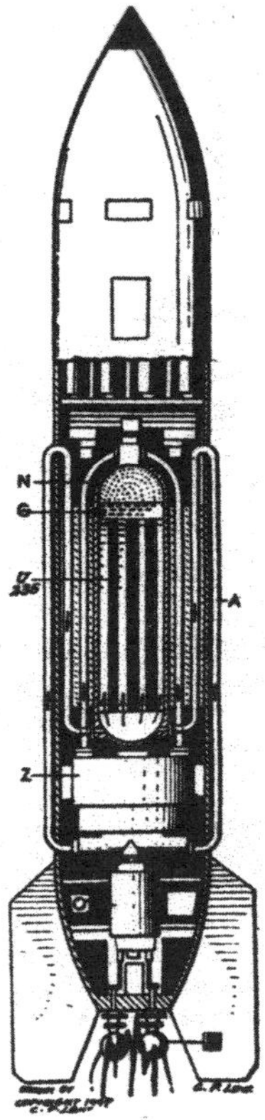

The ATOMIC ROCKET OF THE FUTURE. The drawing above shows a proposed design for an Atomic Rocket. The rocket is driven through the reaction of bolts of high voltage electric current discharged from the rear. The fuel, in this case, Uranium-235, is housed in elongated tubes which are located within the high pressure boiler N. Heavy water G is made to circulate around the tubes to absorb the heat generated by U-235. The resultant high pressure steam drives the turbine Z which turns the generator O. Bolts of high voltage current produced by the generator are discharged continuously to drive the rocket. The turbine exhaust is condensed while it circulates in tubes A around the outside of the rocket where temperatures are extremely low, and then it is returned back to the boiler N to be used all over again. The turbine can also be driven through mercury vapors or some other liquid.

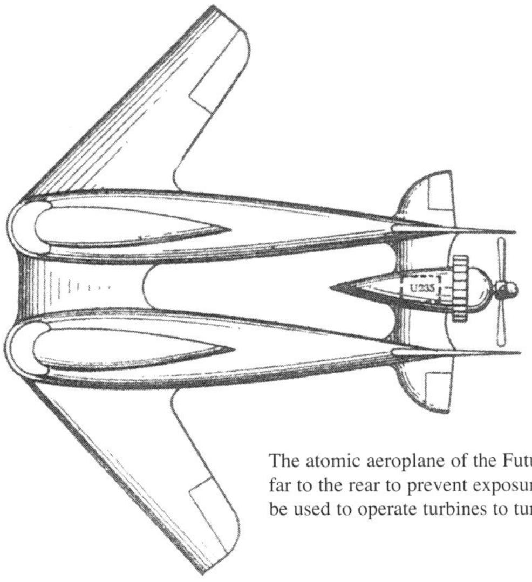

The atomic aeroplane of the Future. The atom motor will be located far to the rear to prevent exposure to radiation. Mercury vapors might be used to operate turbines to turn propellers or operate jets.

The atom car of the future. The atom motor will be located far to the rear and will be shielded by a heavy lead screen from its occupants. Mercury vapors might turn turbines which will be located within the frame of the rear wheels.

X211 Flight Test Program
August 1959 Edward Test Center

PROPOSAL

- Modify a Douglas C-133 "Cargomaster" to serve as a Test Bed for the X211 twin unit.
  _ Provide for operating the right engine only.
- Flight envelope maximum Mach number0.62
  maximum altitude 35,000 feet.
- Fly 1 year from go-ahead.
- Test 75 hours first year
  100 hours second and subsequent.
- Cost: of modification..........$1.50 mil.
  of instrument operation .....$1.25 mil./year.

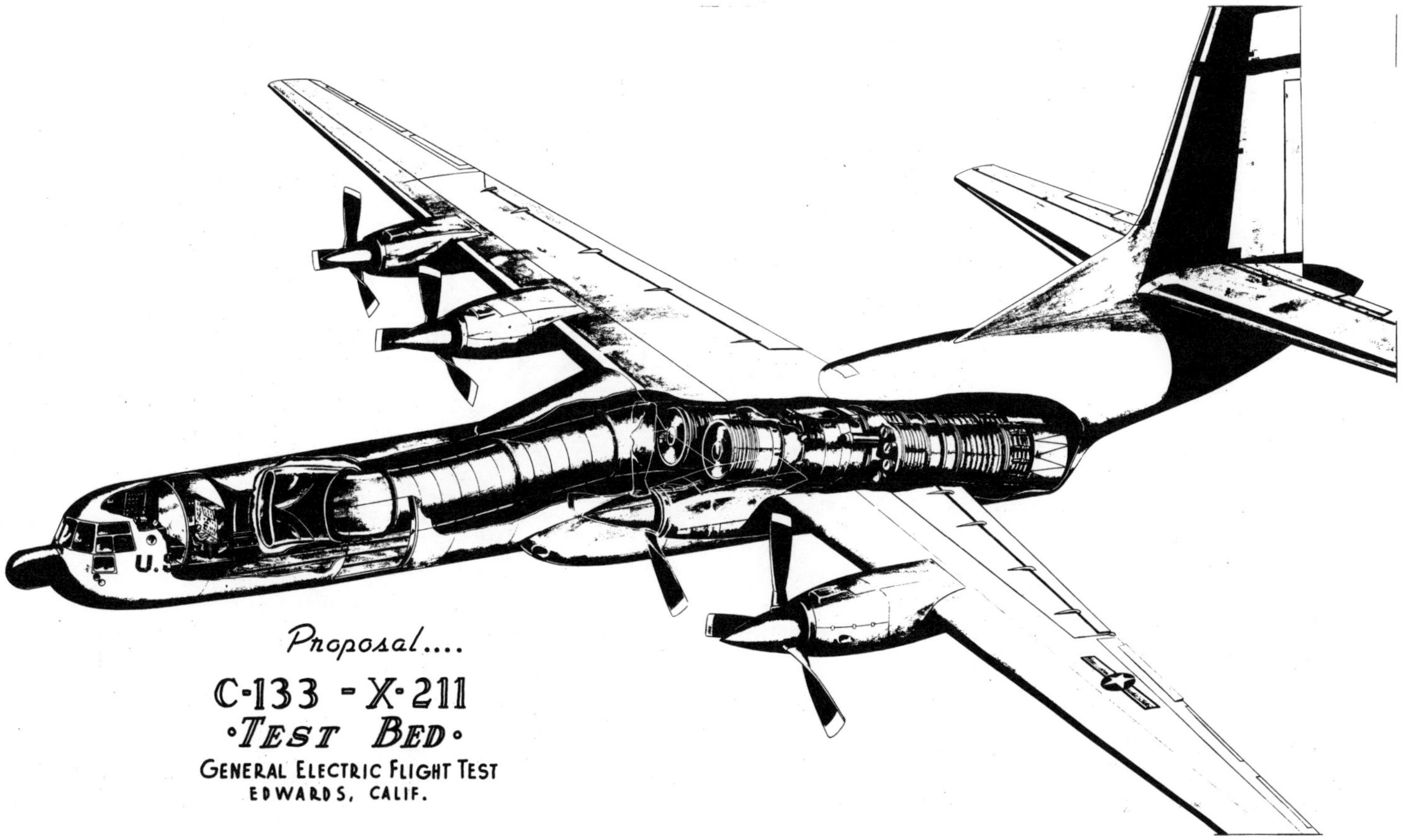

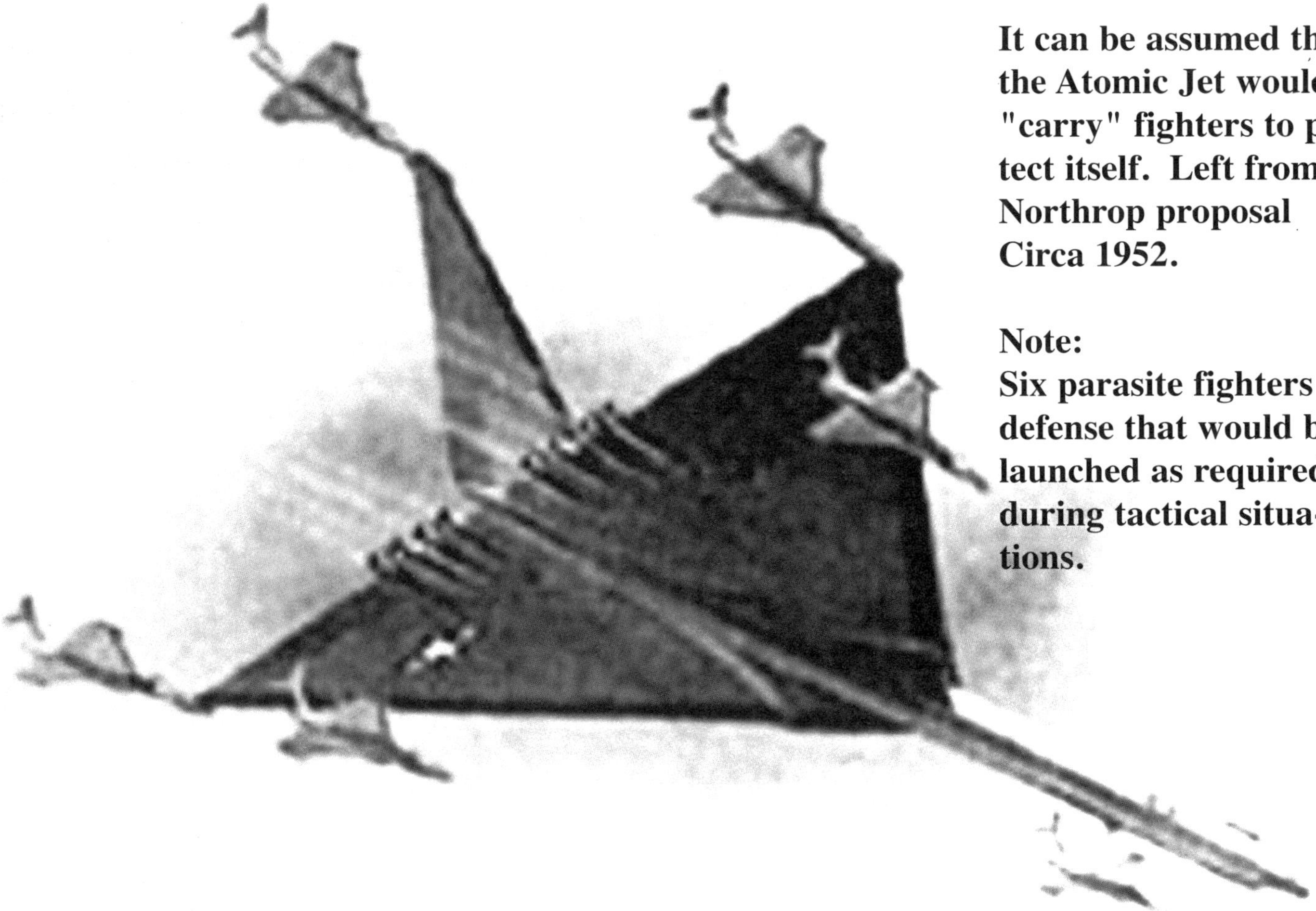

It can be assumed that the Atomic Jet would "carry" fighters to protect itself. Left from a Northrop proposal Circa 1952.

Note:
Six parasite fighters for defense that would be launched as required during tactical situations.

Two early stage proposals for Nuclear Jet Propulsion Circa 1950s.

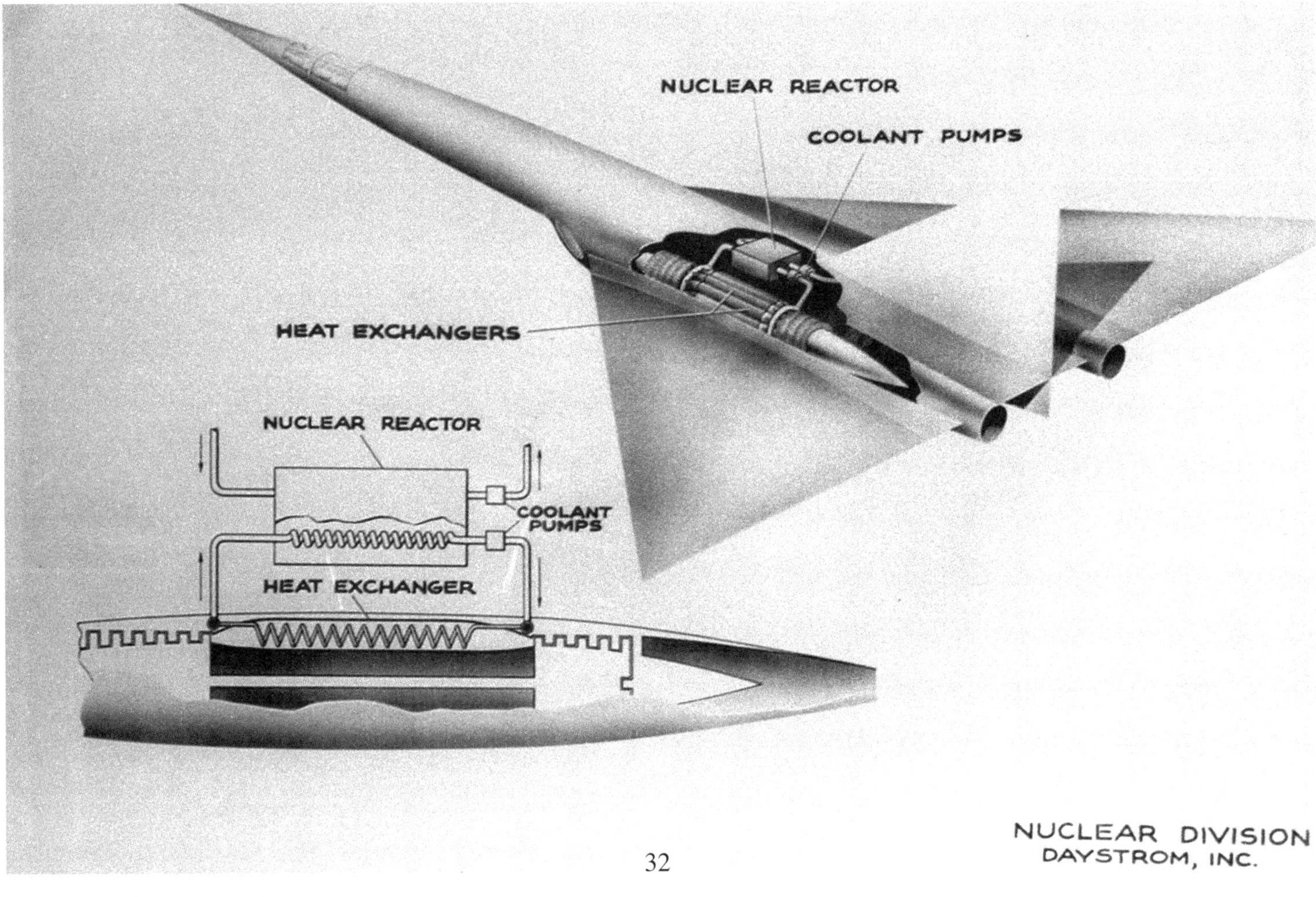

Proposed Convair NX-2 with Indirect cycle engines

Canard configuration was chosen for the first nuclear-powered aircraft because it allowed the crew's compartment to be located a maximum distance (over 100 ft.) from the reactor and the aircraft's center of gravity.

Smithsonian National Air and Space Museum Photo 00-21135

First Nuclear Plane Details Shown in Convair Design
First U.S. nuclear-powered aircraft, to be built for the Air Force by Convair Ft. Worth, will have a subsonic, canard configuration and will be about the same size as late model B-52's, weighing around 450,000 lbs. Model on these pages shows the general arrangement of the aircraft, which can accommodate either General Electric direct-cycle or Pratt & Whitney indirect-cycle nuclear engines without major modifications. Present Air Force schedule calls for extensive flight test using conventional engines with hydrocarbon fuels before the first flight on nuclear power late in 1965. This schedule is based on continuation of the current expenditure rate of about &150 million per year and no major, unforeseen technical difficulties.

Convair press release (1959)

Note: This configuration above of the NX-2 is different from the A/C on this books cover. This design would use an indirect cycle atomic engine.

NX-2 for direct-cycle engines.

Smithsonian National Air and Space Museum Photo 00-21085

To the right is the NX-2 direct-cycle design for the GE Series of nuclear propulsion engines.

Developments during latter years

During 1956, the GE HTRE -1 reactor was successfully tested in Idaho. (See chapter on Design of the Reactor / Engine). This direct- cycle, water- moderated reactor with Nichrome- clad fuel elements was used successfully to power a J47 turbojet engine. GE-ANP Department personnel were now at 1,700. In June, USAF General Curtis LeMay told the Joint Committee that he was interested in achieving nuclear flight at the earliest practical date. General Keirn said ground tests were possible in 1959, and first flight of a prototype in 1960.

However, revised Fiscal Year 1957 funding resulted in an 18 month slippage. A policy decision cut back funding for the Aircraft Nuclear Program.

In December 1956, GE's HTRE-2 test reactor operated successfully at the Idaho Test Station. In the following two months of testing, over 150 hours of nuclear -powered turbojet engine operations were completed. A meeting of the President, DoD. and Budget Bureau in December planned to eliminated the P&W indirect cycle program.This was not accomplished, however. They also reduced the effort on the more mature GE direct cycle program.

In January of 1957 a USAF Scientific Advisory Board recommended less emphasis on engine and airframe development, and more on research and development. In February, the Littlewood Committee of the DoD. began a review of the Aircraft Nuclear Program . They urged efforts to achieve an early flight for the prototype nuclear system. Defense Deputy Secretary Quarles said no flight date would be set until the propulsion system was developed adequately for a military bomber. The Joint Committee expressed concern about the lack of firm program objectives and direction. The DoD then appointed a panel of generals to review the program and missions contemplated. The Joint Committee urged the DoD. to proceed with a vigorous ANP. They met again with Secretary Quarles to emphasize their concern. He agreed and proposed a plan for first flight in 1960.

To the right is HTRE-3. Looking into the reactor core. Note: The fueled cartridges in the center. The reactor was 51 inches in diameter with a 35-inch core length. Total weight was 20 tons. It produced up to 30 megawatts of power.

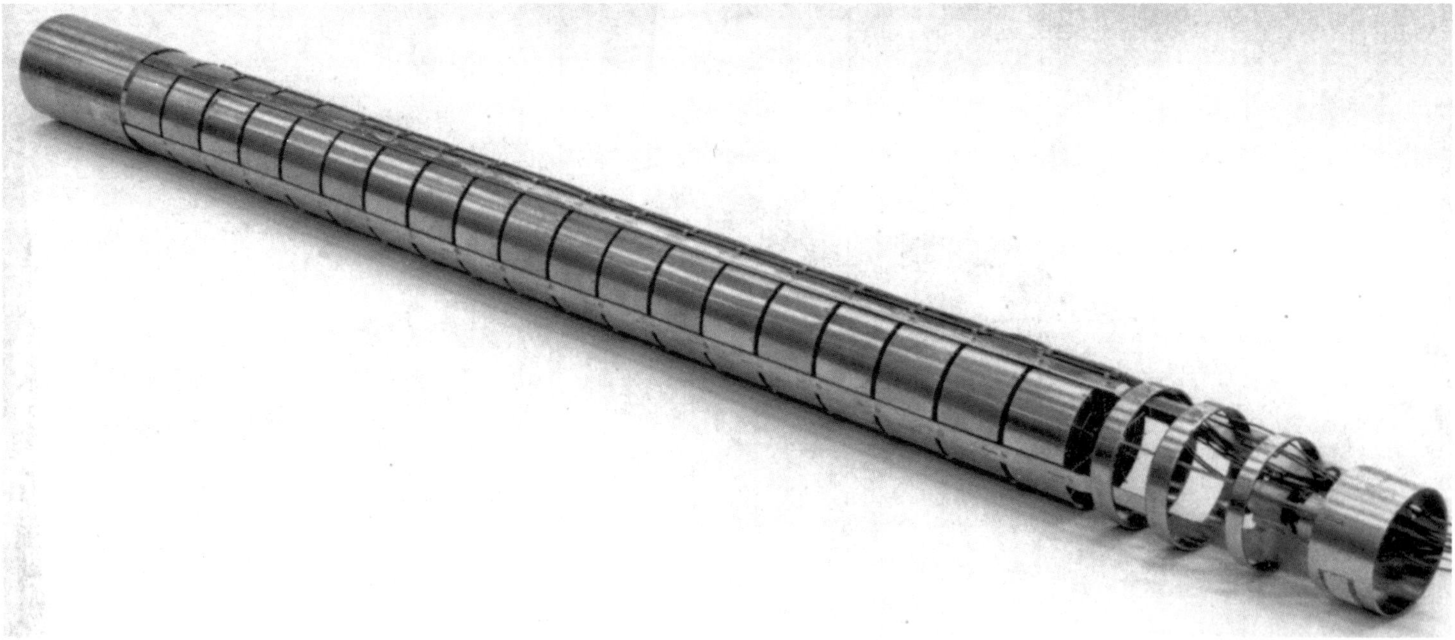

Above is a fuel cartridge, one of 37 in the reactor core. Each cartridge consisted of many sub inserts stacked together. Compressor air would flow thru the multi concentric ring arrangement. Picking up additional energy from the reactor process.

GEAE Photo

HTRE-3 Reactor Core weight: 20 Tons output: exceeding 30 megawatts

GEAE Photo

Meanwhile, GE-ANP Department personnel were now at 2,900. A test of ceramic tubular fuel elements in the GE HTRE-2 reactor showed that water vapor corrosion of beryllium oxide in the fuel elements was a major problem. Note: this problem was later solved by a process of co-extruding the fuel elements with a coating of yttrium oxide and zirconium oxide.

In May of 1957, the Budget Bureau sent a directive to executive agencies requiring that the FY 1959 budget be held the same as or lower than in FY 1958. A congressional committee recommended early fabrication and flight testing of a prototype system in the 1960's. Budgetary ceilings cause a slippage in the time schedule for ground test of the direct cycle system. P&W efforts on the indirect cycle are reduced to a low level.

On October 4, 1957, the USSR launched Sputnik, the world's first satellite. Congress sent a letter to the president recommending a speed-up in the Nuclear Aircraft Program. In December, the Hunsaker Committee expresses concern over potential hazards of nuclear flight over land. They recommended flight tests be made from an island or coastal base.

In 1958, President Eisenhower requested his science advisor, Dr. Killiam, to review the Aircraft Nuclear Program. The recommendation was for greater emphasis on advanced materials capable of producing a higher- performance reactor. The early flight proposal of the USAF was postponed. By December, GE's HTRE-3 reactor began tests in Idaho. The HTRE-3 was a development test assembly consisting of controls, shielding, a direct cycle reactor and two modified J47 jet engines with dual combustion chambers. All this was mounted on a railroad car for quick access into and out of the test cell. The reactor was 51 inches in diameter, with a 35- inch core length. Total weight of the reactor assembly was 20 tons. During the shakedown test run, a power excursion melted some fuel elements. This resulted in a six-month delay while repairs were made.

In December, Aviation Week Magazine broke the news that the Soviet Union had successfully tested a nuclear- powered bomber. Drawings were included in the article. Newspaper editorials criticized the administration for foot-dragging, and Congress called for more funds to meet the Soviet threat. U.S. Intelligence agencies were highly skeptical, and Eisenhower said there was no reliable evidence of the Soviet flights. He was correct - the Soviets had not flown a prototype nuclear- powered bomber. To our knowledge, they never even built one much less flew one.

In February, 1959, Secretary Quarles briefed a congressional committee and stated the Aircraft Nuclear Program had to remain oriented toward development of a nuclear propulsion system, rather than production of an aircraft, until the material problems had been solved. That same month, Convair was selected as the prototype airframe contractor. In April, Secretary Quarles took congressmen on the Armed Services Committee to GE's Evendale, Ohio, plant for a briefing on the program. Congressmen believed that Quarles was influenced positively by what he saw. Unfortunately, Secretary Quarles died of a heart attack in May. Congressional hearings were postponed indefinitely. Thomas Gates succeeded Quarles, but since he had no knowledge of the ANP, he turned over management to Dr. Herbert York, a high DoD official. York prepared a new report on the ANP. In June of 1959, York stated that the ANP should be re-oriented toward development of more advanced materials, and that greater emphasis should be placed on the indirect cycle. All target dates for nuclear flight were eliminated. Another change of strategy !

In August 1959, General. Keirn retired from his role as chief of the ANP Office. D. R. Shoults left the GE ANP Program, and was replaced by Sam Levine. The following month, the GE HTRE-3 reactor became critical at the Idaho station and initially operated at ten megawatts, meeting or exceeding all expectations. By December, the HTRE-3 was endurance tested at full power (31.8 megawatts) for 126 hours. Maximum fuel element temperature was 2030 F. All components were in excellent condition. Both the reactor and jet engine performed in accordance with all design specifications.

In 1960, Dave Shaw became manager of the GE-ANP Department. In May, the Flight Engine Test Facility was completed at the Idaho Test Station. This two- acre building provided for operation of prototype flight worthy nuclear power plants mounted in an air frame mock-up. Also included in this facility were a "hot shop" with remote master-slave manipulators, and a radioactive materials laboratory.

In November, Dr. York, Director of Defense R & D, concluded that the nuclear- powered flight Program did not measure up in competition for money and man-power. There was no longer a need for nuclear powered aircraft in view of the development of ICBM's, nuclear submarines (with ballistic missiles), mid-air refueling of SAC bombers, and the many air bases surrounding the Soviet Union.

In January of 1961, President Kennedy's administration began. Robert McNamara was appointed Secretary of Defense. The Aircraft Nuclear Program received his personal attention, and with the advice of Dr. York, he concluded that one of the budgetary items that could be trimmed at no cost to the nation's security was the nuclear airplane. In March, the AEC directed all contractors to discontinue any work related to the nuclear-powered aircraft in view of President Kennedy's decision to cancel the development program for a nuclear powered military bomber. In May of 1961, the Aircraft Nuclear Program was terminated. In 15 years, over one billion of 1950s dollars was spent on research and development.

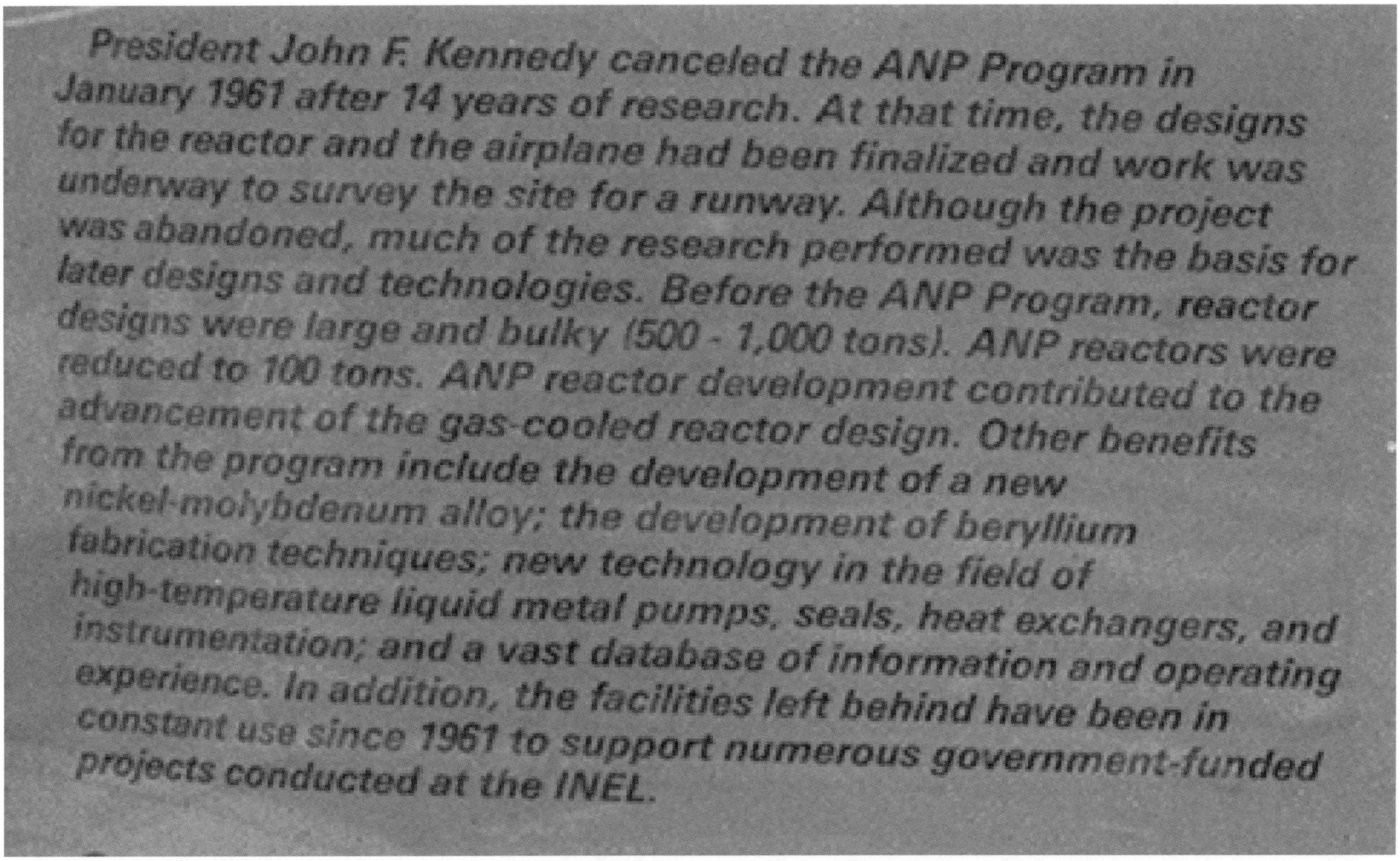

This is a photo of the information board that was placed in front of the two nuclear test rigs in 1985 at the Arco, Idaho test site.

The Idaho Nuclear Test Station

In the southern part of Idaho, there is a vast open landscape that once was a lava flow region. This harsh land is very dry and barren. In 1953, the Atomic Energy Commission began to develop this area, 35 miles west of Idaho Falls.

The first two buildings constructed were the Cold Shop, a traditional machine shop. Next to it was the Hot Shop, a building that would handle radioactive materials. Operators worked behind seven- foot -thick concrete walls, and six inch foot- thick windows, operating with remote-control manipulators, slings, and tools to handle the "hot" materials. An engine test facility was constructed to test jet engines using chemicals (jet fuel) for power. Also constructed were a Low Power Test facility, and a Shield Test facility. One and a half miles away, the Reactor Core Test facility was built. The GE- designed HTRE series reactors were tested here. This building would be the site of nuclear power testing, with the jet engines brought to the building on railroad tracks. They brought the jet engines from other on-site facilities. A shielded locomotive delivered the engines, on their own platform, to the Core Test facility, then backed away. An underground control room was nearby, containing engineers and test equipment, to obtain 500 pieces of data per run. It was "state of the art" for the 1950's. The control room was entered through a 450- foot tunnel. A huge smokestack was used to expel the hot gases from the engine. This effluent contained radioactive materials, so the testing could only be done when wind conditions permitted.

A later building was the Flight Engine Test facility. This was a huge building that could hold a nuclear- powered aircraft. The size was 300 by 320 feet, and it contained shielded walls to accommodate a "hot" aircraft. It was to be used during the bomber prototype phase that never came to be. A 15,000 foot long runway was also planned within the complex but never built.

Below is the remote test site at Arco, Idaho as seen today. Because of the nature of nuclear testing this site provided containment to any possible radioactive emissions.

INEEL Photo

A member of the GE ANP team returned to the Idaho Test Site recently. The "hot" shop is still being used by the AEC, and remote manipulator devices are still in use. The test stand structures housing the HTRE-2 and HTRE-3 nuclear reactors are still on site, although they have been moved about ten miles, at a remote location. Descriptive information panels tell the story of how the GE reactors produced the first jet engines powered by nuclear power.

Building the freestanding hanger for the NX-2 320x300 ft. 1953.

INEEL Photo

Completed hanger. The 15,000 runway was never built.

INEEL Photo

Front view of the NX-2 hanger. 320x320 ft. Note: Man standing in center.

INEEL Photo

Special shielded locomotive built by GE to move the HTRE series reactors into and out of the test site.
Note the thick glass viewing port to protect the driver from radiation hazards.

INEEL Photo

"Hot" shop for disassembly of HTRE series reactors.
Note: Twin rail system to distribute the weight of the reactor/engine assembly which exceeded 600,000 lbs.

INEEL Photo

Inside the newly completed "hot" shop at the Idaho Test Facility.
Note: Turntable bottom left

INEEL Photo

Progress in the field of atomic energy depends, to a great extent, upon man's ability to handle materials "hot" with radioactivity. Shown here is the world's largest mechanical hand designed for such work. It can lift, with equal ease, a 7000-lb load or a fragile egg, "lending a hand" where no man could possibly survive.

Sheet metal test building for HTRE series testing.
Note: Exhaust tube leading to exhaust stack. The sheet metal building could be moved back and forth depending on weather conditions. Wind conditions could sometime limit the days that testing could be allowed.

INEEL Photo

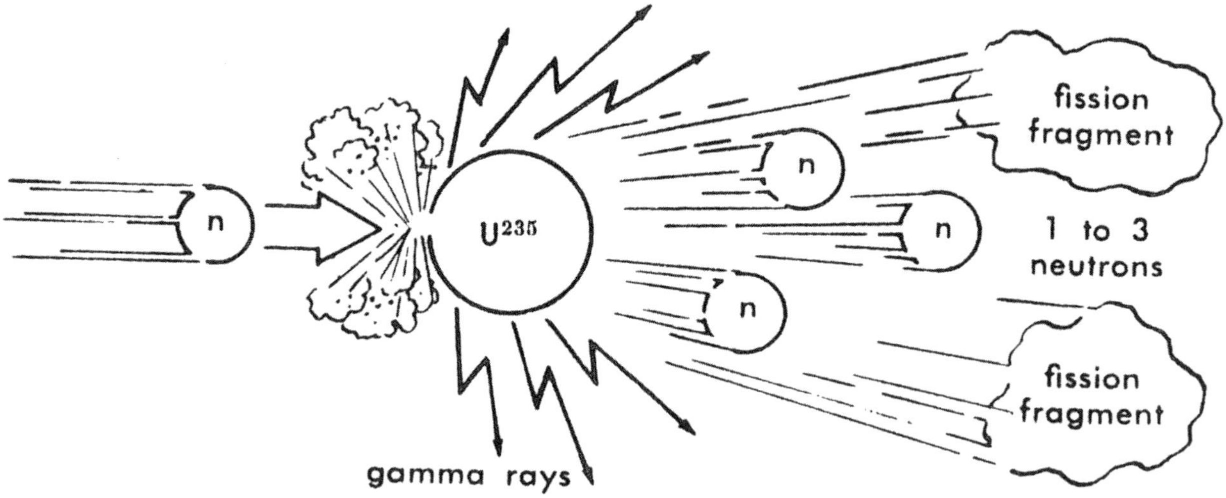

The fission reaction of uranium-235, which is the source of heat and radiation that powered the X211 series of engines.

The Beetle built for remote handling of the radioactive engine parts.

INEEL Photo

Right. The "Beetle." This manned, shielded vehicle was designed for use in the hangar. Its purpose was to remove the reactor power plant from the aircraft, and eventually, from an actual aircraft. It was never used in Idaho, but moved to the nuclear rocket program.

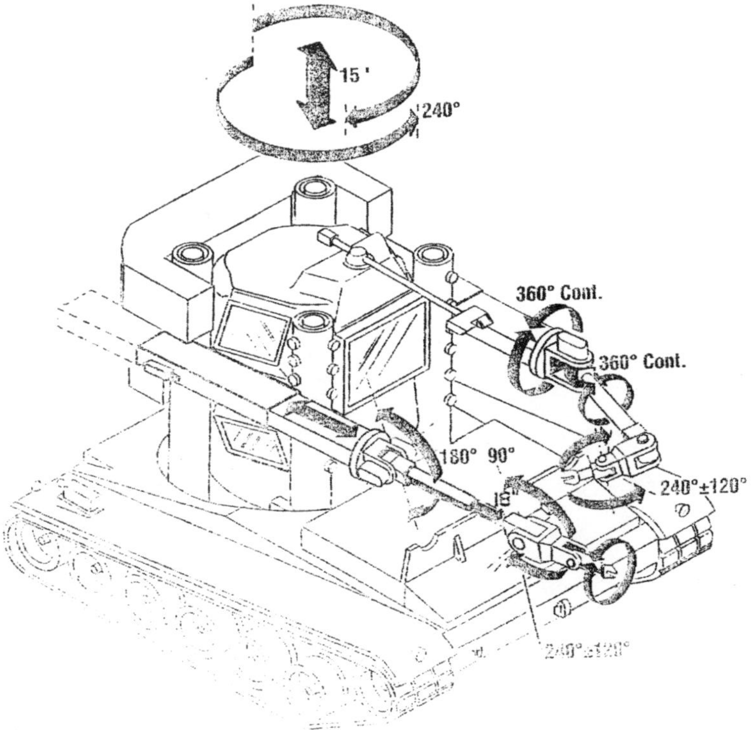

Beetle pursuing other interests !

INEEL Photo

The Russian Nuclear Program

As early as 1950, the Soviet Union had "**plans**" to construct a nuclear- powered aircraft. One design was a flying boat, with a weight of two million pounds. Four nuclear-powered turboprop engines would power this huge machine. Wing span was proposed to be 400 feet. Five layers of shielding were proposed to protect the crew and passengers this aircraft was proposed as a 1,000 - passenger commercial liner. The Russians studied both the direct cycle and indirect cycle methods of obtaining power for their jet turbines. The Soviet studies envisioned about 160,000 pound of shielding to protect the crew. This was very similar to the weight required for the P1 reactor / power plant that GE had designed for the NB-36H. These developments, as far as we know, never were pursued into hardware for flight.

In 1957, the launch of Sputnik not only started a race for space, but also a technology battle between the free world and the Communist world. Congressmen wrote to President Eisenhower urging him to speed up work on the Nuclear Aircraft Program. USAF General Keirn, who headed up the ANP Office, Said: "I"m sure each of you is aware of and appreciates the seriousness of any potential threat to our seacoast military installations, plus industrial and population centers, posed by a large enemy submarine fleet. Imagine in addition to this a fleet of enemy high speed aircraft continuously patrolling the airspace just outside the early-warning net, capable of air-launching a devastating missile attack, followed by high-speed penetration or attack of our hardened installations."

A year later, Aviation Week Magazine published an article stating that the Soviet Union had test flown a nuclear- powered aircraft. Quote: "A nuclear powered bomber is being flight tested in the Soviet Union. Completed about six months ago, this aircraft has been flying in the Moscow area for at least two months. It has been observed both in flight and on the ground by a wide variety of foreign observers, from Communist and non-Communist countries." Drawings were included, with dimensions (195 feet long), weights (300,000 pounds), and powerplant details. The aircraft was said to have contained two large turbojets powered by nuclear reactors, plus two conventional turbojets in separate pods.

Senator Russell of Georgia said in a TV interview: "The report that the Russians have test flown an atomic powered aircraft is an ominous new threat to world peace, and yet another blow to the prestige and security of our nation and the free world. It follows in tragic sequence to the Russian success of last Fall in launching the world's first artificial satellite. If the report is true, it means that we are today faced with a new weapon of terrifying consequence."

Today it is obvious that no such nuclear aircraft ever existed in the Soviet Union. According to Doctor Herbert York: "Stories to that effect were simply one more clear and very obvious attempt at what may be called self-serving intelligence. Those of us who had all the facts in the matter and who knew there was no real basis for any of these claims were hamstrung in any attempts we made by the secrecy which always surrounds real intelligence information".

On the following several pages are period reports from 1958-1959. These reports give insight into what industry and government knew regarding the Soviet Unions plans-progress and dreams for nuclear powered flight. The first article from (Aircraft Missiles, December 1959) came out after the December 1, 1958 Aviation Week report but appears to have been a basis for the article. An additional report (date unclear) called (Soviet Nuclear Propulsion by R.G. Perelman) is not published here. However, it is by in large the same as published here.

Russia's nuclear aircraft

many configurations studied, design now beyond theoretical boundaries; prototype construction underway

This article is a condensed version of a Russian manuscript entitled, "Atomic Energy in Aviation." Author is Yu. N. Sushkov. Translation was done by General Electric Co., Aircraft Nuclear Propulsion Dept.

Aircraft and Missiles • December 1959

When discussing the application of atomic energy in aviation, several questions arise:

- Are atomic aircraft engines necessary at all?
- Do they have any advantages over ordinary engines?
- Is it possible to build reliable and nonhazardous atomic engines for aircraft?

Only by using nuclear fuel can the problem of long flight-range at supersonic speeds be solved.

Nuclear fuel consumption would be 2 million times smaller than the consumption of kerosene in conventional aircraft. An aircraft weighing 100 to 150 tons would consume only 500 grams of uranium-235 in a round-the-world flight at 2000 km/hr.

The Basis of Nuclear Power

Nuclear power is obtained by fission of nuclei in isotopes of some heavy elements. These isotopes are: uranium-235 found in natural uranium in the amount of 0.7 per cent; artificially-produced uranium-233; and plutonium-239.

This is how the fission reaction proceeds in uranium-235. A free or so-called "wandering" neutron impacts a nucleus of uranium-235 and imparts energy to it. The nucleus becomes excited and unstable. In most cases it fractures. At the moment of fracture 2 to 3 new neutrons are released. These are called secondary neutrons. If the piece of uranium is of sufficient size, each of the secondary neutrons can impact another uranium nucleus and cause new fission. This process continues, and under controlled conditions the fission reaction can be self-sustained. Reactions of this type are called chain reactions.

Speed of the fission fragments, immediately after fracture, is equal to 10,000 to 15,000 km/sec. The fragments impart energy to the surrounding atoms, increasing the speed of their random movement. This increases the temperature and nuclear energy is released in the form of thermal energy.

Fig. 1. General view and schematic of an aircraft with two atomic turbojet engines: (1) inlet cone, (2) compressor, (3) reactor, (4) control rod, (5) turbine, (6) jet nozzle.

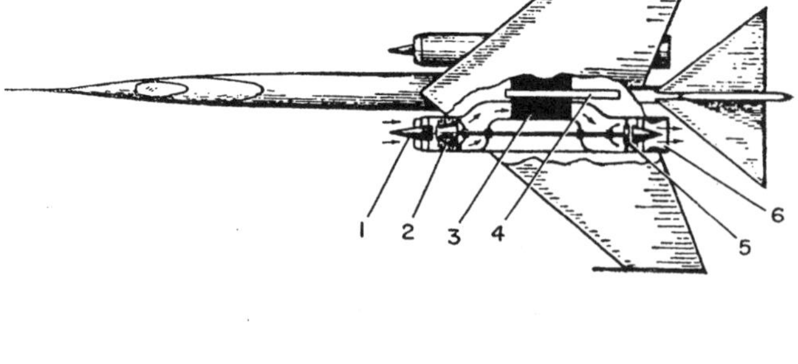

Therefore, all contemporary atomic-power installations utilize heat engines—steam or gas turbines.

The Direct Cycle System

Fig. 1 shows a schematic of an airplane with two atomic turbojet engines operating on the direct-cycle principle. The reactor replaces the combustion chamber. The air which is heated in the reactor is returned to the turbines. After passing through the turbines, the air is ejected out through the jet nozzle at high speed to create thrust.

Inlet cones, shown on the schematic, are placed in front of the engine to decrease the intensity of the shock waves which occur at supersonic speeds; shock waves decrease propulsive efficiency.

Supplying air to the reactor requires tremendous ducts and makes the arrangement of atomic powerplants on an aircraft inescapably difficult. In order to heat the air in the reactor, a large heat-exchange surface area is required: about 800 to 1000 sq meters for an aircraft with a flying weight of 100 to 120 tons. Thus, the reactor is forced to large dimensions. This of course, results in an increase of aircraft weight and a decrease in its flying speed for a given power.

Indirect Cycle

Heating the air in an atomic turbojet engine may also be accomplished indirectly in a heat exchanger, with the aid of a metallic heat-transfer agent (Fig. 2). The heat-transfer surface required to heat the liquid metal is many times smaller than the surface required to heat the air directly. The reactor can therefore be smaller and lighter. But the weight of the heat exchanger and the heat-transfer medium add to the weight of the engine. Whether an engine with the air heated directly in the reactor, or with the air heated indirectly in a heat exchanger, is more advantageous depends on the flying weight of the aircraft and its designed flight speed. The indirect system is especially favorable in multi-engine aircraft, since the operation of more than one engine is possible from a single reactor.

Increasing Thrust

Part of the energy of the heated air is lost in the turbine; therefore, the air temperature behind the turbine is 150 to 200°C lower than in front of the turbine. The air exit speed from the jet nozzle, and accordingly the thrust, depends on the air temperature in front of the nozzle. The higher the air temperature in front of the nozzle, the higher is the propulsive thrust. Therefore, it is advantageous to heat the air additionally behind the turbine.

The simplest way to do this is to place an additional heat-exchanger between the turbine and the jet nozzle. The heat-transfer medium from the reactor first passes into the additional heat-exchanger, gives up some of its energy to the air flowing into the nozzle, then returns to the original heat-exchanger to heat the air just being directed towards the turbine.

In this system the coefficient of useful turbine action is lowered and the nuclear-fuel consumption is increased. But, since consumption of nuclear fuel is very small, lowering the turbine efficiency is made up by an increase in propulsive thrust.

Another way to increase the air temperature in front of the jet nozzle is to use a steam or gas turbine instead of a compressor. This turbine would not use the thermal energy from the air heat exchanger; therefore, the velocity of the air exhaust from the nozzle and consequently the propulsive thrust would increase.

Fig. 2. Schematic of an atomic turbojet engine with liquid-metal heat-exchanger: (1) reactor, (2) control rod, (3) liquid-metal pump, (4) heat exchanger, (5) inlet cone, (6) compressor, (7) jet nozzle.

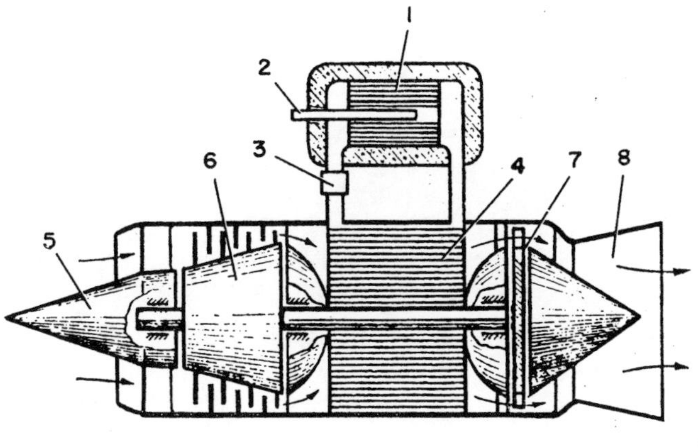

Fig. 3. Schematic and general view of an aircraft with two atomic compressor-drive engines: (1) inlet cone, (2) air compressor, (3) gas turbine, (4) gas compressor, (5) heat exchanger, (6) jet nozzle, (7) nuclear reactor, (8) control rod.

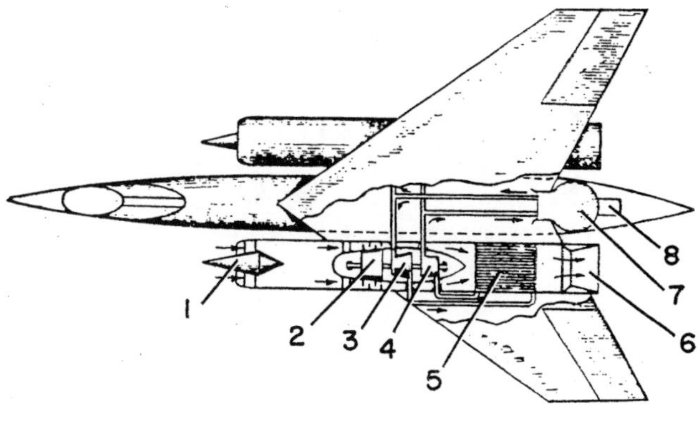

Fig. 4. Schematic of an atomic turboprop engine with mercury turbine: (1) propellers, (2) reduction gear, (3) air compressor, (4) mercury turbine, (5) reactor, (6) control rod, (7) mercury pump, (8) condenser, (9) jet nozzle.

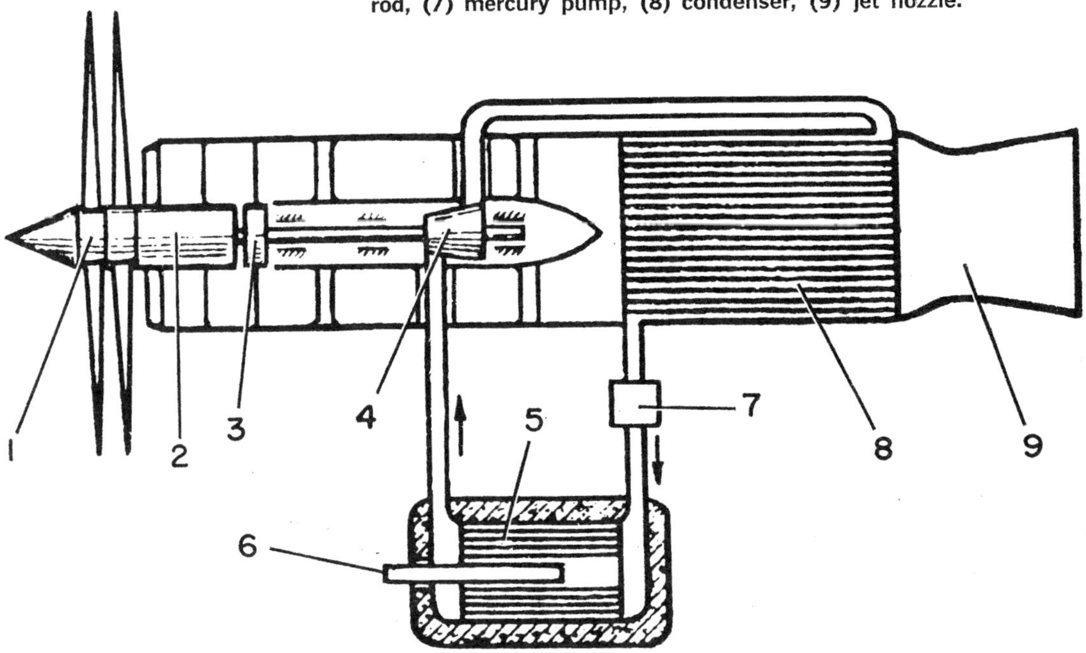

Fig. 3 is a schematic of an aircraft having two atomic engines with the air compressor driven by a helium turbine.

The power ratio is approximately as follows: If the power of the helium turbine is 150,000 hp, then 100,000 hp is used by the helium compressor and 50,000 hp by the air compressor. The efficiency of such an arrangement is not high.

Mercury-Vapor Turbine

Power requirements are less when the air compressor is operated by a mercury-vapor turbine. Mercury is changed into vapor in the reactor, then is directed to the turbine which is connected to the air compressor by means of a shaft. The vapor flows from the turbine to the condenser-heat-exchanger. Upon condensing into the liquid mercury, the vapor heats the air. Because of the low temperature of condensation of mercury, even at the increased pressure in the condenser, the propulsive thrust will be somewhat less than in the case of a helium turbine. However, the nuclear-fuel requirements will be lower, because, instead of the 150,000-hp turbine, a turbine of only 55,000 hp will be required. As before, 50,000 hp will be used by the air compressor, but only 5000 hp will be used by the mercury pump. If the coefficient of useful action of the steam turbine is the same as for the gas turbine, then the decrease in steam-turbine power-requirements will also result in a decrease in reactor power, for the same propulsive thrust. But as a rule, the coefficient of useful action of a gas turbine is much greater than for a steam turbine. Therefore, the economic gain from the decrease of steam-turbine power is lost to a significant degree, and the resulting reactor-power for both engines is about the same. When comparing the drive of an air compressor by gas and mercury turbines, it should be kept in mind that it is much more complicated to build a reliable reactor where mercury is changed into vapor than it is to build a gas-cooled reactor.

The Turbine Bypass

There is one more way of increasing the air temperature in front of the nozzle. The idea is to pass only a part of the air of the atomic turbojet engine through the turbine. By doing this, the thermal energy of the air flowing through the turbine will be considerably lowered, and it will not be able to play an effective role in the creation of propulsive thrust. However, that part of the air which enters the exhaust nozzle after by-passing the turbine will have

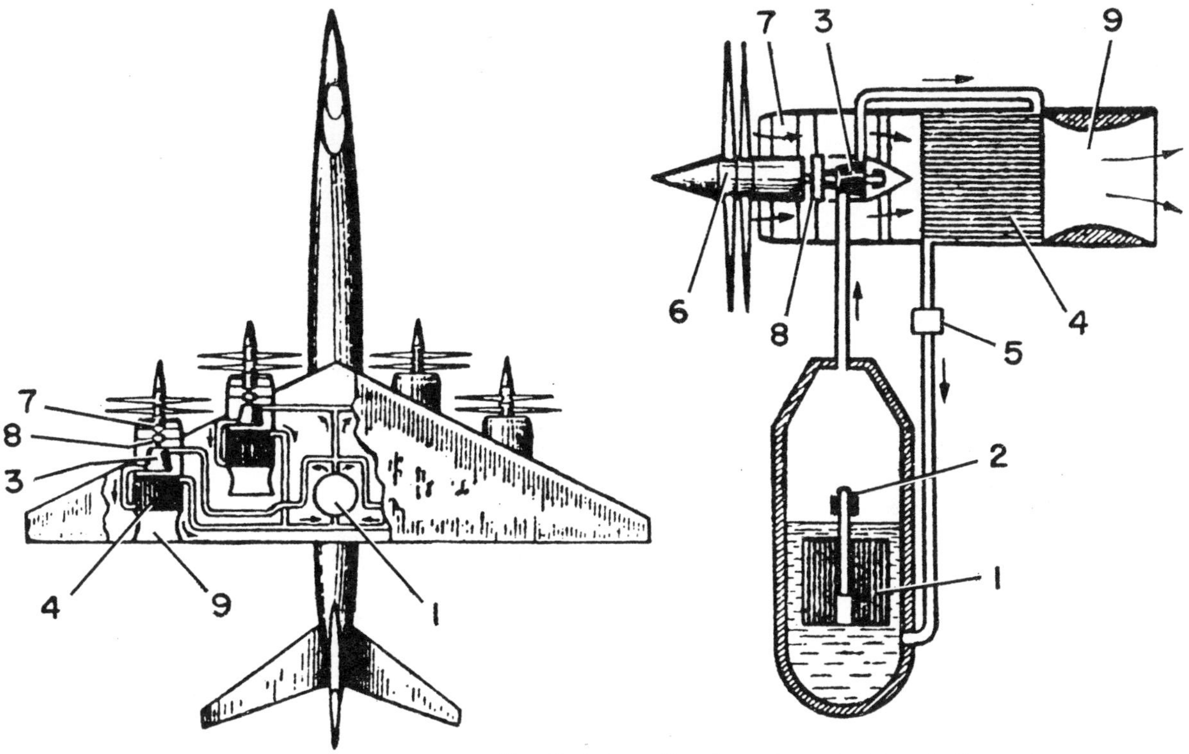

Fig. 5. Schematic of an aircraft with a boiling-water reactor and steam-turbine engines: (1) boiling-water reactor, (2) control rod, (3) steam turbine, (4) condenser, (5) water pump, (6) propellers, (7) reduction gear, (8) air compressor, (9) jet nozzle.

the same temperature as air in the reactor or the heat exchanger. Therefore, the velocity of the exhaust from the jet nozzle will be greater than in the case where all the air flows through the turbine. Thus, the propulsive thrust will increase.

The Turboprop

An atomic turboprop-engine differs from an atomic turbojet only by the presence of a propeller and reduction gears. Since the turbine of such an engine must be much more powerful than the turbine of a turbojet, it should be multistage. It is possible to utilize steam or gas turbines in the atomic turboprop. An air compressor is required only to force air through the condenser heat-exchanger. Fig. 4 shows the schematic construction of an atomic turboprop engine with a mercury turbine.

A "boiling water"-type reactor is shown on an aircraft which has steam-turbine atomic turboprop engines in Fig. 5. This type reactor consists of a thick-walled pressure-vessel containing ordinary water. A grid of enriched uranium rods, encased in zirconium, is placed in the water. The evolved heat converts the water into steam. A steam separator is located either in the upper part of the reactor or above the reactor. It separates small water-particles from the steam and returns them to the reactor. The steam is ducted to the engines' steam turbines. After being used, the steam passes into a condenser, is cooled by a stream of air, is condensed to water, and is again pumped into the reactor. The turbine drives the propellers through reduction gears.

The Ramjet

The application of atomic energy in other kinds of aircraft engines is also possible. Fig. 6 is a schematic of an airplane having two atomic ramjets located above the fuselage, and one atomic rocket-engine located in the fuselage. The ramjet engine is essentially a "flying reactor" to which an inlet diffuser, which directs a supply of air to the reactor, is attached at the front. Behind, there is a jet nozzle through which hot air is discharged at a greater velocity than the velocity of the airplane. Propulsive thrust is created because of the difference between the inlet and exhaust velocities of the air. The fault of a ramjet engine, either atomic or ordinary, is that it can increase its propulsive thrust only at high speeds. Therefore, an aircraft must have some other engine besides the ramjet in order to take off and accelerate.

Russia's nuclear aircraft...

continued

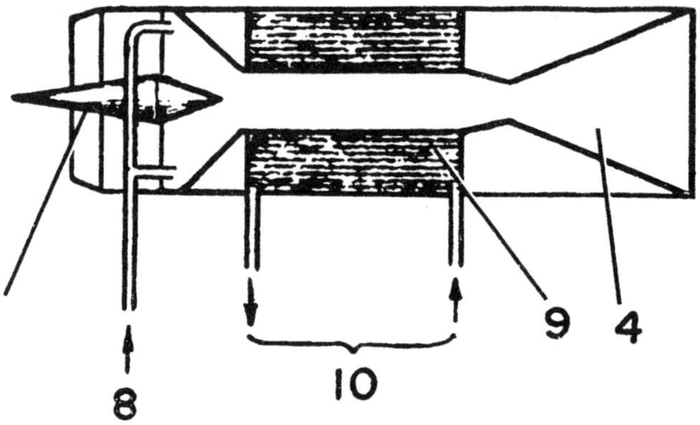

Fig. 6. General configuration and schematic of an aircraft with two atomic ramjet and one atomic rocket-engine: (1) inlet cone, (2) reactor for ramjet engine, (3) control rod, (4) jet nozzle, (5) reactor for rocket engine, (6) pump, (7) hydrogen tank, (8) ducts for gaseous nuclear fuel, (9) moderator, (10) reactor coolant.

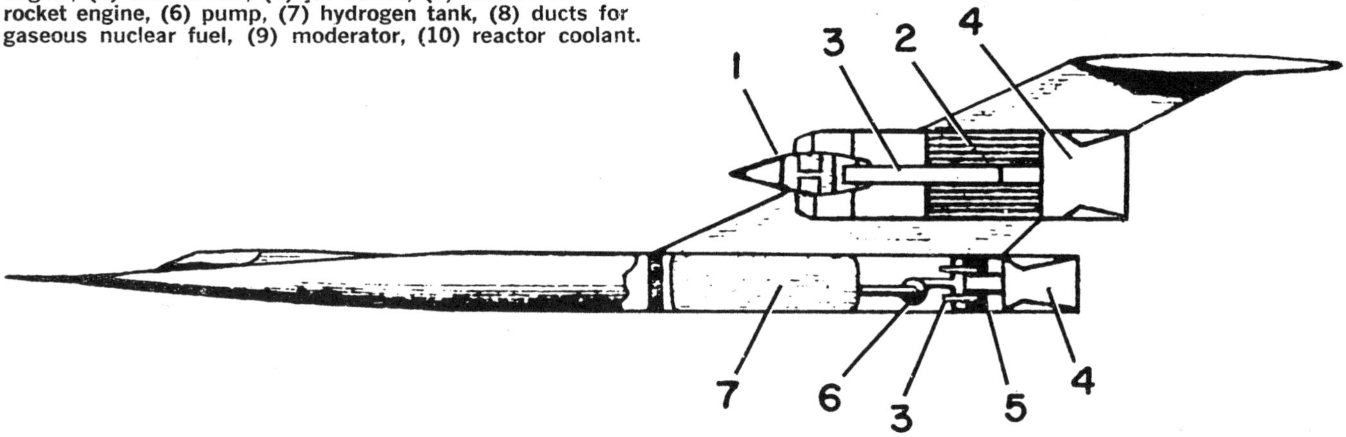

If an atomic ramjet engine should operate on gaseous nuclear fuel, the air which is mixed with this gaseous fuel is directed into the reactor cavity. There, the nuclear reaction takes place and as a result the air is heated. Then the air passes on out through the nozzle. The reactor in this engine is a powerful source of neutrons. Naturally, this is accompanied by the liberation of heat. This heat must be removed from the reactor and can be used for the operation of the turbine on any other engine installed on the aircraft. The possibility of using this type engine depends on the quantity of gaseous nuclear fuel that is consumed. Calculations show that at neutron flux densities in the reactor which are possible at the present time, the probability of neutron collision with the gaseous nuclear fuel nuclei is very small. To achieve the required heating of the air in the engine, the consumption of gaseous nuclear-fuel would be very large. Therefore, it is not practical to build such an engine.

Development Status

Schemes for aircraft atomic-engines described here, of course, do not cover all variations possible; however, they show some of the possible design concepts.

It is possible to build an atomic powered airplane. The atomic engines, the reactor, and protective shield would not weigh more than certain conventional aircraft with full fuel. Minimum flying-weight of an aircraft for which these conditions hold, would probably be 90 to 100 tons. The payload would be 5 to 10 tons, and the aircraft would thus be able to transport 50 to 100 passengers.

When should the first flight of an atomic airplane be expected. What are the practical achievements in the areas of application of atomic engines in aviation?

Progress on the construction of an atomic airplane is beyond purely theoretical boundaries and has passed into the experimental phase, construction development, and experimentation with separate components. We are at the threshold of the next phase—the construction of an experimental prototype.

✦

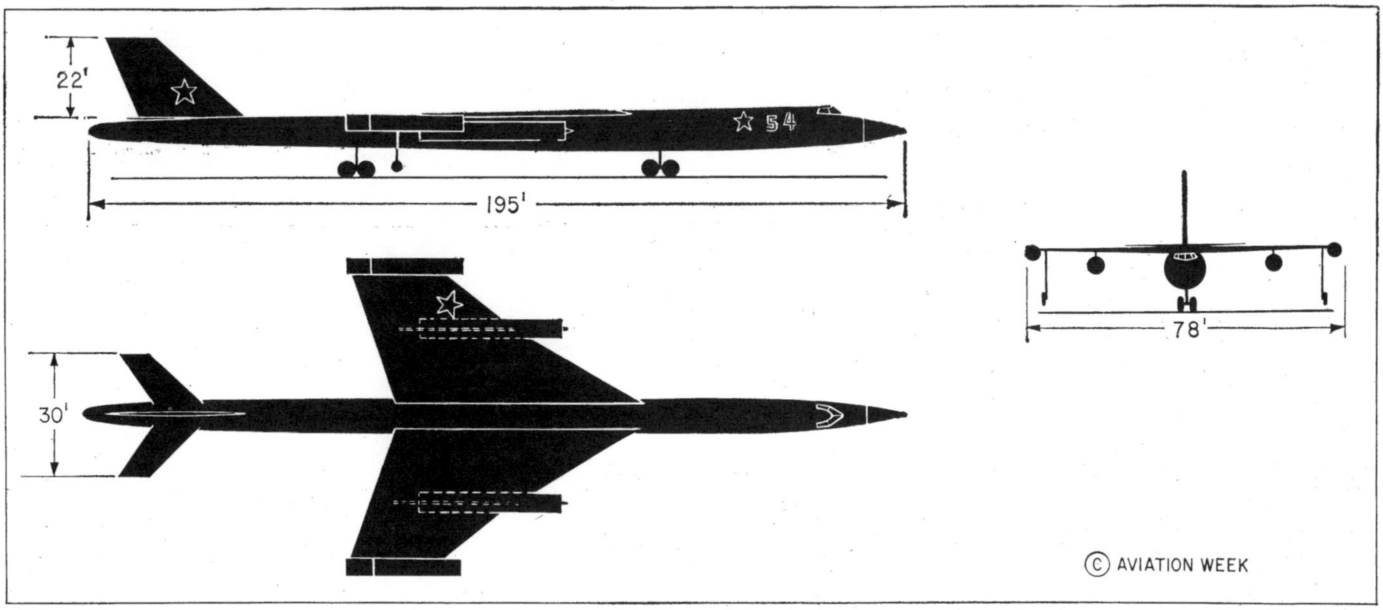

AVIATION WEEK ARTIST'S conception of how Soviet nuclear-powered bomber prototype looks in three-view. Note bicycle main landing gear with outriggers folding up into wing from wingtip and tin high-speed airfoil. Nuclear powerplant air intakes are approximately 6-ft. in diameter. Radiation shielding details are not available, but divided shielding concept is used with heavy shielding around nuclear powerplants and additional shielding protecting crew compartments.

Soviets Flight Testing Nuclear Bomber

Atomic powerplants producing 70,000 lb. thrust are combined with turbojets for initial operations.

Washington—A nuclear-powered bomber is being flight tested in the Soviet Union.

Completed about six months ago, this aircraft has been flying in the Moscow area for at least two months. It has been observed both in flight and on the ground by a wide variety of foreign observers from Communist and non-Communist countries.

In its initial flight testing, the new aircraft is powered by a combination of nuclear and conventional turbojet engines. Two direct air cycle nuclear powerplants are housed in 36-ft.-long nacelles slung on short pylons about midway out on each wing. These nuclear powerplants, with 6-ft.-diameter air intakes and using small but high power reactors to replace the combustion chambers in the turbojet cycle, produce about 70,000 lb. thrust each.

They are supplemented by two conventional turbojets installed in wingtip pods fitted with short afterburners to provide about 35,000 lb. thrust each for takeoff performance. The conventional, chemically fueled turbojets are used primarily for safety purposes during the early flight test program of the nuclear powerplants. In later versions of the aircraft, they may be retained for high-speed dash performance or replaced by two more nuclear powerplants after their reliability has been proved in flight.

The Russian nuclear-powered bomber is not a flying test bed in the sense that earlier U. S. Air Force and Navy programs had called for installing a nuclear powerplant in a conventional airframe such as the B-36 or Saunders-Roe Princess flying boat solely for test purposes. The Soviet aircraft is prototype of a design to perform a military mission as a continuous airborne alert warning system and missile launching platform similar to the USAF CAMAL project for which Convair and Lockheed are now making design studies (AW Nov. 10, p. 37). The CAMAL mission was recently described in detail by Maj. Gen. Donald Keirn (see box page 28).

In its present configuration with both nuclear and conventional turbojets, the Soviet aircraft has a performance capability in the high subsonic and low supersonic speed ranges with its range limited only by engine component life and crew endurance.

The Soviet nuclear-powered plane has a fuselage about 195 ft. long and a 78 ft. wingspan. The delta-type wing is sweptback on both leading and trailing edges. From an initial angle of 60 deg. sweepback at the wing root, the leading edge changes to about 55 deg. sweep at the inboard engine pylon mounts and beyond to the wingtips to produce a "cranked" effect familiar on British bombers such as the Handley Page Victor and Avro Vulcan. Trailing edge of the wing is swept about 15 deg. This delta-type wing uses a relatively thin, high-speed airfoil confirming eventual performance goals for this design in the Mach 2 speed area.

Vertical tail rises about 22 ft. above the fuselage. It is a typical "sail" type fin used by Soviet designers to ensure good directional stability. Horizontal tail surfaces have a span of 30 ft. and are swept back at about the same angle as the outboard wing panels. They apparently are kept well clear of the nuclear powerplant efflux both by placement high on the fuselage and by span length.

Aircraft has a gross weight of about 300,000 lb., and a wing loading of about 118 lb. per sq. ft.

The direct air cycle nuclear powerplant has been described in some detail in Soviet technical publications (see diagram on page 28). In a text published last year by the Military Press of the Soviet Defense Ministry entitled "Application of Atomic Engines in Aviation," the direct air cycle powerplant is described as follows:

"The simplest is a design that differs from the ordinary turbojet engine only in that the combustion chamber is replaced by a reactor . . .

"This simplest design permits obtaining the highest specific performance parameters. In this case, the air duct

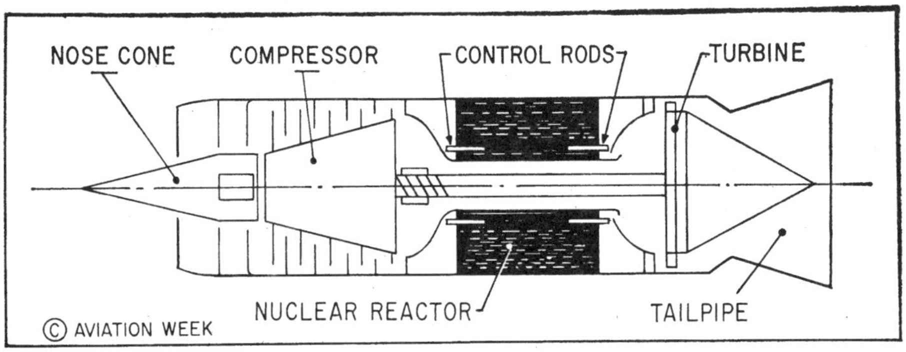

SCHEMATIC DIAGRAMS from Soviet technical source on a direct air cycle atomic turbojet. Nuclear reactor is used in place of normal chemical fuel combustion chamber with air and compressor shaft passing directly through reactor.

becomes a uniflow duct where the airflow through the engine is at all times parallel to the engine axis in a straight line so that hydraulic resistance is at a minimum.

"The air is heated directly in the reactor without an intermediate heat transfer agent. This simplifies the design and eliminates excessive heat loss. However, this design which is simple in principle, is exceedingly difficult to realize. The shaft connecting the turbine with the compressor has to pass through the reactor. Cooling the shaft under these conditions becomes a difficult, and actually the key, problem.

"The point is that the shaft not only becomes heated as a result of heat transfer from the hot reactor parts but considerable liberation of heat occurs within the shaft itself due to scattering and absorption of neutrons and gamma rays by the shaft material. So much heat is liberated in the shaft that cooling of the shaft changes from a simple engineering matter to a complex problem whose solution will govern the very possibility of developing an atomic turbojet engine on the basis of this 'simplest' design."

This direct air cycle nuclear powerplant represents the same approach pursued by the Aircraft Nuclear Propulsion Department of General Electric Co. under USAF and Atomic Energy Commission sponsorship since 1951 at facilities in Evendale, Ohio, and Arco, Idaho.

D. R. Shoults, general manager of the GE nuclear propulsion program, reported that actual operational tests at the ARCO facility had "proved the feasibility of a direct air cycle aircraft propulsion system and demonstrated its performance." The Shoults report was made in a paper prepared for the Second United Nations International Conference on Peaceful Uses of Atomic Energy held last September in Geneva, Switzerland (AW Sept. 22, p. 55).

He described the results of operating heat transfer reactor experiment No. 1 (HTRE-1) during more than 100 hr. of turbojet running on nuclear power without "any failures of any sort".

Shoults also reported that HTRE-1 determined the following:

- **Integrated performance of** the reactor and turbojet engine in the powerplant system.
- **That overheating in portions of** the reactor would not lead to local flow starvation and progressive overheating.
- **Integrity and life of key** components of the system.
- **Ability to carry out extensive** remote handling of radioactive components.
- **Control response of the reactor** and its relationship to turbojet engine control.
- **Unanticipated problems and** their possible solutions.

The approach to flight testing nuclear-powered aircraft both in this country and in the Soviet Union calls for initial operations on conventional chemical fuels to prove out aircraft systems and familiarize the crew with operational techniques. However, even in operational use of a nuclear-powered military aircraft, nuclear powerplants may be operated initially on chemical fuel until the reactor temperatures are sufficiently high for full power operation. Then chemical fuel combustion can be phased out and the turbojet permitted to operate on nuclear power alone. Similarly, on return from a mission the reactors would be shut down some distance from destination, with the return to base and landing again made on chemical fuels. For this type of operation, a chemical combustion system also must be incorporated in the nuclear powerplant package.

Powerplants Flight Tested

Although much of the early flight testing of the Soviet nuclear aircraft has been conducted on conventional fuel the nuclear powerplants have definitely been tested in the air. Fission of one pound of uranium 235—most frequently mentioned in Soviet technical literature along with plutonium 239 as an airborne reactor fuel—will liberate about the same amount of energy as the burning of 1,700,000 lb. of gasoline.

There is no specific information available on the types of shielding employed on the new nuclear-powered aircraft but recent Soviet technical literature has been studded with brief but positive references to a major "breakthrough" in shielding techniques. Soviet technical literature emphasizes the concept of divided shielding with heavy use of stainless steel in the engine and aircraft structure to provide containment shielding for neutron radiation and another type of shielding protecting the crew quarters from gamma radiation. The extreme length of the aircraft fuselage also would be aimed at maximum separation of crew from the radioactive engines.

The podded installation is best suited

Nuclear Plane's Military Mission

Washington—"Imagine a fleet of 'enemy' high speed aircraft continuously patrolling the airspace just outside our early warning net capable of air launching a devastating missile attack against our hardened installations. Through a consideration of these capabilities, combined with those possessed by the intercontinental range ballistic missile, the degree of possible future threat of surprise attack immediately becomes apparent. . . .

"An ideal airborne alert manned aircraft system must carry a large payload and remain on nomadic patrol for extended periods of time in various areas of the world. It must maintain continuous communication with appropriate headquarters and be capable of instantaneous reaction with air launched missiles. When required, the system should be capable of following up the missile launching phase with a low-level high-speed penetration of the enemy's heartland in order to seek out and destroy hardened targets or targets whose locations are not sufficiently well known to permit attack by long-range missiles.

"The combination of these features can best be achieved through the application of nuclear propulsion. . . . Such a system may be similar in weight and size to the B-52 and be capable of carrying a heavy payload on extended endurance mission. Because of its endurance, varying armament load and high speed capability at minimum altitudes, its operational versatility would be outstanding. . . . But perhaps even more important is its inherent operational flexibility for meeting various limited war and peacetime situations."—Maj. Gen. Donald J. Keirn USAF deputy chief of staff for development for nuclear weapons.

U. S. Nuclear Powered Aircraft Program

• 1946—Nuclear Energy Propulsion for Aircraft program (NEPA) organized under Air Force contract with Fairchild Engine & Aircraft Co. project manager at Atomic Energy Commission's Oak Ridge, Tenn., laboratory. Purpose is to study feasibility of applying nuclear power to aircraft and develop components for such a system. Consultants from aircraft engine manufacturers and universities particpate.
• 1948—Lexington committee, primarily Massachusetts Institute of Technology professors, called in by AEC to study NEPA program. Recommend continuance of project.
• 1949—Oak Ridge AEC laboratory establishes nuclear aircraft propulsion research program.
• 1949—Technical Advisory Board reviews NEPA program, recommends continuance.
• 1950—Fairchild top management change and company is relieved of NEPA project.
• 1950—Divided shielding concept developed and NEPA project again approved by a review technical committee.
• 1951—USAF concludes NEPA had demonstrated feasibility of nuclear propulsion of aircraft.
• 1951—May, General Electric organizes aircraft nuclear propulsion program as part of its Aircraft Gas Turbine Division under USAF and AEC contract.
• 1951—May, Pratt & Whitney gets USAF contract to explore closed cycle nuclear powerplant.
• 1951—Fall, GE gets approval for development of direct air cycle type nuclear powerplant.
• 1952—Spring, GE begins work to develop experimental nuclear powerplant to be tested in B-36 with flight date projected for 1956. GE aircraft nuclear propulsion project becomes a separate department and begins work on six-day week.
• 1953—Spring, Convair makes first test aircraft equipment in radiation environment.
• 1953—May, Pratt & Whitney Aircraft gets AEC backup contract on USAF closed cycle work.
• 1953—Spring, Charles E. Wilson, Secretary of Defense, decides to cancel Convair-GE aircraft nuclear propulsion program. GE program retained by USAF Secretary Harold Talbott's diversion of unallocated funds to project.
• 1953—Fall, ground test reactor, first of its type outside AEC, designed and built by Convair and put into operation.
• 1954—USAF cancels B-36 nuclear engine test program. AEC continues support of GE program on limited scale. GE program with HTRE-1 goes back to system development with X-39 engine aimed at high-performance nuclear powerplant system to be operational in 1956.
• 1954—Fall, aircraft shielding test reactor, first flying reactor built and designed by Convair, put into operation.
• 1954—USAF develops WS-125A nuclear bomber program aimed at high subsonic cruise bomber with supersonic dash capability. Pratt & Whitney and General Electric are engine contractors and Convair and Lockheed airframe contractors.
• 1955—Navy begins nuclear-powered seaplane studies.
• 1955—September, first operation of a reactor in the air aboard a B-36 modified by Convair.
• 1955—October, engine-airframe company teams for system WS-125A selected by Air Force. GE and Convair as one team, Pratt & Whitney and Lockheed the other.
• 1956—January, GE successfully rates X39 turbojet engine with nuclear reactor in HTRE-1 for successful operation on nuclear power.
• 1956—Spring, USAF plans construction of 10 kilowatt test reactor for aircraft materiels testing at Wright Field, Ohio.
• 1956—September, National Advisory Committee for Aeronautics begins construction of nuclear test reactor at Lewis Flight Propulsion Laboratory, Cleveland.
• 1956—September to January, 1957, GE operates HTRE-1 with turbojet engine running 100 hr. on nuclear power without failure of any kind.
• 1956—Fall, USAF cancels WS-125A program. Powerplant development continued with no specific aircraft goals.
• 1957—May, Pratt & Whitney completes Connecticut Aircraft Nuclear Engineering Laboratory (CANEL) at Middletown, Conn., under AEC contract.
• 1957—June, Soviet Union ground tests nuclear aircraft powerplant system.
• 1957—August, USAF cancels Pratt & Whitney closed cycle engine development contract. CANEL cuts operations to greatly reduced scale. AEC continues support of P&WA program on small scale.
• 1957—Fall, General Electric proposes accelerated program aimed at getting flying test bed with nuclear powerplant into air as soon as possible.
• 1958—March, President Eisenhower informs Congress there is no urgency in nuclear aircraft propulsion program, rejects accelerated program and authorizes continued low budget development program.
• 1958—June, USAF proposes CAMAL nuclear-powered aircraft program to develop continuous airborne alert, missile launching and low-level penetration mission.
• 1958—July, Navy gets funds to conduct feasibility study of a nuclear-powered aircraft using Pratt & Whitney engines.
• 1958—August, first flights of Soviet nuclear-powered bomber prototype are observed in Moscow region.
• 1958—December, Air Research and Development Command scheduled to analyze Lockheed and Convair CAMAL design studies preliminary to possible award of prototype construction contract.

to the direct air cycle type nuclear powerplant since its operation makes the entire turbojet engine radioactive. The pod would facilitate engine removal by remote control for ground radiological safety and make replacement of the powerplants relatively simple.

Taking a nuclear-powered military aircraft from the early flight test stage, through which the Soviet aircraft is now passing, to a fully operational capability for both airborne early warning and missile launching capability probably will require at least 18-to-24 months. A nuclear powered aircraft requires extensive testing for other flight operational subsystems other than the powerplants and their operation under varying degrees and different types of radioactivity.

The current Soviet milestone in testing nuclear powerplants in flight in a military prototype is the result of a high-priority and financially well supported program stretching back through nearly eight years of research and development.

High priority for the nuclear aircraft program was assigned during the current Sixth five-year program which began in 1956 and will end in 1960. During the past few years, Soviet technical and popular publications began a steady crescendo in their coverage of atomic aircraft powerplant developments in addition to marine atomic powerplants for icebreakers, submarines and surface vessels.

This similar type of publication build-up has preceded every new major Soviet technical achievement, including the intercontinental ballistic missile and the Sputniks.

As long as a year ago, there were brief but specific mentions in the Soviet technical press of successful ground testing of atomic aircraft powerplants. Recent speculative stories in the Soviet popular press suggest conditioning the Russian people to an announcement of a spectacular achievement by an atomic-powered aircraft in the near future, probably a nonstop, non-refueled flight around the world.

AVIATION WEEK ARTIST'S conception of Soviet nuclear-powered bomber shows large nuclear powerpiants suspended from pods midway under delta wing; conventional turbojets with short takeoff afterburners on wingtips, and 195-ft. fuselage to aid in radiation protection.

26 AVIATION WEEK, December 1, 1958

The program that never was (Authors note)

From Aviation Week
Dec. 1, 1958

The Soviet Nuclear-Powered Bomber

On page 27 of this issue we are publishing the first account of the Soviet nuclear-powered bomber prototype along with engineering sketches in as much detail as available data permits.

Appearance of this nuclear-powered military prototype comes as a sickening shock to the many dedicated U. S. Air Force and naval aviation officers, Atomic Energy Commission technicians and industry engineers who have been working doggedly on our own nuclear aircraft propulsion program despite financial starvation, scientific scoffing and top-level indifference. For, once again, the Soviets have beaten us needlessly to a significant technical punch.

While this Soviet achievement is a truly remarkable feat, it is not beyond the technical state of the art in our own nuclear aircraft propulsion program. The difference lies rather in the top priority and steadfast support accorded the Soviet program by its top political leadership and the technical timidity, penny-pinching and lack of vision that have characterized our own political leaders' attitude toward the goal of nuclear-powered aircraft for both military and civil purposes.

This is a story that has become all too familiar to Americans in recent years, punctuated by the Soviet triumphs with the first medium-range ballistic missile in production and military deployment; the first successful ICBM firings followed by an ICBM production rate that is now more than 15 per month, and the trio of Sputniks. This also could be the epitaph carved on the tombstone of this country's genuine technical development capability if we continue much longer on this course.

We are sure that there will be the usual chorus of good gray voices from high official places attempting to "pooh pooh" the existence of a Soviet nuclear-powered bomber prototype and coining smooth weasel-worded phrases to deprecate its significance even if its existence is finally admitted, as finally it must be. For the basic facts on this Soviet aircraft are known in official circles both here and abroad.

The credibility of the same "gray voices" has, of course, diminished in recent years because they used the same tone and phrases to soothe the country before the appearance of Soviet nuclear weapons, intercontinental bombers, supersonic fighters, medium-range missiles, the ICBM and the Sputnik.

Maj. Gen. Donald Keirn, chief of the USAF-AEC aircraft nuclear propulsion program, virtually let the cat out of the official bag in the question period after a speech in Washington last month by admitting that it would not surprise him if the Soviets flew a nuclear-powered aircraft before the end of 1958. This was like placing a bet on a horse race after you have watched it finish through binoculars and have found a bookmaker who doesn't have a phone.

With an acute awareness that the first world-wide demonstration of the Soviet nuclear plane would generate a major political blast, Gen. Keirn also pointed the finger of responsibility for our own slow progress at the sources who are really at fault. These are the anonymous scientists headed by James Killian, scientific advisor to the President, who operate under a heavy veil of official secrecy and only last spring vetoed a military-industry proposal to accelerate nuclear aircraft development on the basis of promising technical achievement.

Although not mentioned by Gen. Keirn, former Secretary of Defense Charles E. Wilson also must answer for his jeering characterization in 1957 of an atomic-powered plane as a "shitepoke," a bird that has a long neck, big body and can fly, but not very fast, and for his 1953 attempt to wipe out the entire aircraft nuclear propulsion program by cutting off its development funds. The late Harold Talbott, then Secretary of the Air Force, circumvented this Wilson order by diverting some "hidden" USAF funds to the program. AEC also maintained its support.

Since word of the Soviet nuclear-powered aircraft began filtering through the Iron Curtain, the pentagon has hastily revived an active program aimed at a nuclear-powered military aircraft known as the CAMAL project. Mission of this aircraft is described by Gen. Keirn on page 28. Think of the political and military significance of even a small fleet of nuclear-powered military aircraft that, as Gen. Keirn described, can cruise indefinitely off the territorial limits of the U. S. maintaining a continuous airborne alert and warning system combined with the capability of quick launching of missiles with sufficient range to penetrate 1,000 mi. or more and following up this attack with a high-speed, low-level penetration well under and around radar defenses to obliterate key targets missiles cannot locate. It takes no military expert to appreciate the value of this apparatus in the hands of an aggressive, ambitious political dictatorship bent on world conquest.

The development of a nuclear-powered military aircraft involves much more than just producing a satisfactory powerplant, and this is a phase in which we lag even farther behind than in engine development. Every time the financial pinch was applied to the nuclear aircraft program, those in charge, of necessity, starved aircraft systems development to keep the powerplant research alive.

Thus, with the Soviets now in the initial flight testing of a nuclear-powered military prototype and the Air Research and Development Command scheduled only this month to make a decision on prototype construction of a similar USAF weapon system, it is clear to even the most conservative technical analysts that we are at least four years behind the Russians in this critical area. Development of such a new and technically radical weapon system to full military capability is a long, painstaking and failure-studded process on both sides of the Iron Curtain. But with such a clear-cut lead, we can expect the Soviets to exploit their nuclear-powered aircraft for political warfare long before it has developed a sound military capability. There already are indications that a nonstop, non-refueled flight several times around the world is being planned by the Soviets with this type aircraft. And, how much political force will an event such as this impact on our allies, the neutrals and our enemies?

During the past few years, we have heard much from our political leaders on how much we can or cannot "afford" for the defense of this country.

These were the same years that we have been belabored with vigorous efforts to cut the strength of our military forces in being and jeopardize our military future by sabre slashes through the research and development budget.

These were the same years the Soviets appeared first with their huge turbojet and turboprop gas turbines, their medium-range ballistic missiles, ICBM and Sputniks.

In view of these Soviet technical achievements, it is more pertinent to ask:

How much longer can we "afford" this kind of leadership and still survive as a free nation?

—**Robert Hotz**

The Human Element

(USAF Study, from 1960)

In space cabin simulator studies performed at the School of Aviation Medicine, a number of seven day flights have been conducted, using highly trained, carefully selected US Air Force subjects. These pilot subjects were committed as an integral component of a man-machine system, and have been exposed to relevant conditions of a closed environment of a nuclear crew compartment. These conditions include a simulated altitude of one half an atmosphere with a gaseous environment equivalent to that of ground level; an extreme degree of physical confinement, including restricted mobility; isolation and sensory deprivation; an abnormal schedule of work and rest - four hours on, and four hours off during the entire seven days; high noise level; and limited facilities for personal hygiene. It is interesting to note that these experienced pilots did adjust to their work schedule, and that they maintained an extremely high level of proficiency for the entire period. In two instances, they could have continued beyond the seven days of flight. These results strongly suggest that problems of work schedule, day-night cycling, boredom, isolation, nutrition, and limited hygiene can all be satisfactorily controlled.

A difficulty in the development of a nuclear powered aircraft is in providing crew compartment space consistent with crew needs. The aircraft will not provide the space provided in a nuclear submarine, for example. In the work area, the crew members are integrated into a highly coordinated arrangement. The aircraft commander and flight engineer are located side-by-side, as are the navigator and defense director. The co-pilot is forward of the flight engineer, across the aisle from the aircraft commander. This layout provides the commander direct visual and verbal communications with all other crew members. It also minimize interruption of flight continuity when crew members shift duties.

The leisure area is equipped to satisfy every personal need, and requiring very little effort to maintain. In approximately 36 square feet of floor space, it provides clothing storage, hygiene facilities, and sleeping accommodations, as well as an integrated food preparation area, and small mess table. A two cubit foot freezer stores 35 pre-cooked frozen meals. Non frozen foods, such as canned juices and other meals, are stored within the drawer spaces assigned to each crew member. Meal preparation equipment consists of a warming oven, a sandwich grill, and fresh water taps.

Superimposed on these long term flight stresses will be the effect of exposure to ionizing radiation. Fission effects from the nuclear reactor in the aft part of the aircraft are intensely radioactive. This radiation, having a wide range of energy, can and do penetrate matter. They can penetrate the human body without being sensed. By a process of ionization they affect living cells, and can produce serious illness and death. Many radio-biologists in the nuclear aircraft project feel that these levels of radiation exposure may be the prime limiting factor in the operation of an aircraft. It is important, therefore, that the biological hazard of dose-effect and the response to various radiation sources be well understood. This will permit levels of exposure, and thereby limits of operation, on the crew members.

In the design of a nuclear powered aircraft, the engineers, because of weight and cost considerations, desires the minimum of radiation shielding. The radio-biologist, on the other hand, desires to have the minimum radiation hazard, which means maximum shielding between the reactor and the crew. The question is: How much shielding is necessary for adequate protection of the crew? How often can an individual be exposed to this radiation without significant detrimental effects? In answering these questions, it is extremely important to determine the maximum level of radiation exposure to the crew. Further testing and experiments will be defined so that maximum aircraft utilization along with maximum personnel protection. (See menu for five day flight)

Ground Support for Nuclear Aircraft

(HS Hutchins, Convair 1960)

With full shielding around a nuclear reactor it is possible to completely contain the primary radiation. But for military aircraft this would impose impractical shield weights and unacceptable performance. Most designs for military nuclear aircraft reflect a compromise between reactor shield weight, flight performance, and personnel factors. The resulting limitations on the reactor shield bring about the radiation environment that must be contended with in ground maintenance procedures and equipment. The airframe and system components will also become radioactive because of irradiation by neutrons during reactor operation. The radiation from the aircraft decays with time, quite rapidly at first, then progressively slower, finally reaching a low, essentially residual value. Even at its highest, this radiation is many orders of magnitude lower that that from the reactor, and does not offer nearly as serious a problem. In the radiation environment, direct maintenance, tempers with personnel exposure control, is practical, and will permit the use of nearly normal procedures, schedules, and equipment.

Three methods of ground maintenance have received the greatest attention. First, to perform all operations from shielded buildings or vehicles, using remotely controlled manipulators and tools. This approach was rejected, because the time penalties are unacceptable for a military aircraft, operational flexibility is impaired, and the facilities are very expensive. A second method was ground "safing" the reactor through shield augmentation - that is, the placing of additional shielding around the reactor on the ground to attenuate its radiation. This method was rejected because of the size and weight penalties it imposes on aircraft design. The third, and accepted method, involves removal of the reactor as the initial step after each flight. Also, its reinstallation is the final step before take-off. With the reactor removed, the high level radiation environment around the aircraft is eliminated, and the bulk of the maintenance and support activities can be accomplished under safe conditions. This approach was verified during the 1955-1957 B-36H Nuclear Test Aircraft flight program. Ground handling by mobile equipment and maintenance under nearly normal conditions is not only possible but entirely practical.

Recommended Menu for 120-Hour Flight

	breakfast	*lunch*	*dinner*
1st day	eaten at home	apple juice pork steak bread and butter jam pineapple poundcake coffee, tea, cocoa	chicken soup beef pot roast potatoes mixed vegetables bread and butter jelly brownies milk, coffee, tea, cocoa
2nd day	orange juice grilled egg and bacon sandwich milk, coffee, tea, cocoa	tomato soup Swiss steak potatoes and peas bread and butter apple and cheese slice coffee, tea, cocoa	pineapple juice meat and spaghetti bread and butter jelly apricots fruit cake coffee, tea, cocoa
3rd day	tomato juice waffles applesauce sausage honey coffee, tea, milk	chicken soup grilled ham and cheese sandwich cookies nuts and candy milk, coffee, tea, cocoa	apple juice turkey sweet potatoes and Lima beans bread and butter cranberry sauce fruit cocktail cookies coffee, tea, cocoa
4th day	apricot juice grilled egg and bacon sandwich milk, coffee, tea, cocoa	orange juice chicken bread and butter cranberry sauce pears pecan roll coffee, tea, cocoa	tomato soup beefsteak bread and butter ketchup apple date pudding coffee, tea, cocoa
5th day	pineapple juice Spanish omelet sausage sweet roll coffee, tea, cocoa	apricot nectar chicken w/gravy potatoes and corn bread and butter honey brownies milk, coffee, tea, cocoa	tomato juice beef patty potatoes and green beans bread and butter peach pie caramels coffee, tea, cocoa

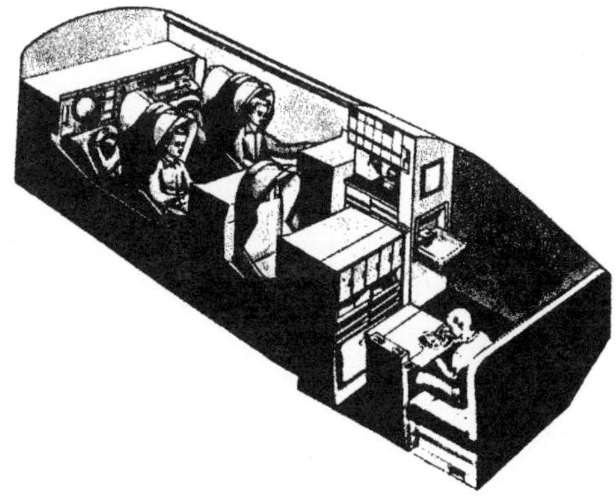

An artist's sketch of the crew compartment design and arrangement the instrumented Nuclear Aircraft Simulation shows th aircraft commander capsule on the far side of the compartment and immediately aft the internal feeding system. Across the aisle is the nuclear engineer, with the bombardier seated behind him. The defense director is seated back-to-back with the craft commander, and the copilot is off duty in the forward leisure.

Reasons that the ANP Program failed

In 1963, the General Accounting Office of Congress did a study of the defunct Aircraft Nuclear Program. The general consensus was that the Program was doomed to failure by the continuing reviews and re-orientations that kept it in a state of flux. Seven basic re-orientations of the program occurred during it's 10 year period an average of one every eighteen months.. The GAO did not comment on the validity of the re-orientations, which varied from a research program, to an accelerated program to develop a weapon system for the military. The GAO did not question whether the ANP Program could have resulted in a successful military program.

"Although it was outside our scope to examine the reasonableness for the frequent changes in the program objectives," the GAO said, "we do not believe that a research and development effort of the complexity and magnitude of the ANP Program can reach its goal in an effective and efficient manner unless a certain degree of stability in objectives is accorded the program".

The GAO counted 13broad reviews of ANP in the last six years of the program. These were to evaluate past accomplishments and to set future objectives. Of the 13 reviews, seven were by the US Air Force, five by the Defense Department, and one by the AEC. The GAO reported that frequent reviews of the ANP Program were made by temporary groupings. The reviews were based on brief visits to contractor plants, plus discussions in Washington DC. "Little continuity could be found among the review groups."

The GAO highlighted eight different ANP Programs during the 1950's:

Flight Demonstration Program - between April 1952 and May 1953

Applied Research & Component Phase - between May 1953 and November, 1954

Weapon System 125-A Program - November. 1954 to December 1956.

Experimental Program - no flights - January to March 1957

Experimental Development Program - April, 1957 to February, 1958

Development Program (military bomber) - March, 1958 to October, 1958

CAMAL Program - (airborne missile launcher) - October 1958 to July 1959

Research & Development Program - July 1959 to program end.

President Kennedy, in announcing the cancellation of the program in a message to Congress in March of 1961, said that "the possibility of achieving a militarily useful aircraft in the foreseeable future is very remote," The cancellation, Kennedy added, will avoid a future expenditure of $1 billion just to achieve first prototype flights.

Other Weapons Systems also a factor

The US Air Force desired to keep the Aircraft Nuclear Program going, because they wanted to continue the development of manned bombers. Most Air Force officers did not want to become "silo sitters" - tending underground intercontinental ballistic missiles (ICBM). In the late 1950's, the Thor, Atlas, and Titan missiles were flying, and were proving their worth. They could carry a thermonuclear weapon and hit within lethal range over intercontinental distances. The Air Force set a higher priority for the ANP aircraft, rather than the ICBM. General LeMay, head of the Strategic Air Command, set top priorities for the B-52 bomber, then the ANP, with long-range missiles at the bottom of the list.

In 1957, with the advent of the Russian Sputnik, a technology race began with the Soviet Union. Missile progress was high on the list, and so was the Aircraft Nuclear Program. In December of 1958, Aviation Week Magazine incorrectly reported the flight of a Soviet nuclear aircraft. This only heightened tensions. Senator Russell of Georgia stated: "the reports that the Russians have test flown an atomic powered aircraft is an ominous new threat to world peace, and yet another blow to the prestige and security of our nation".

President Eisenhower, who of course had all intelligence information available to him, said in December 1958, "There is absolutely no intelligence to back up a report that Russia is flight-testing an atomic-powered aircraft."

By the early 1960's, there were four choices for weapon system development. These included: atomic subs with Polaris missiles, land-based ICBM's, manned bombers (either conventionally powered or nuclear -powered, and tactical bombers and short- range missiles (based in Europe). The

U.S. Government had choices to make they chose to cancel the nuclear bomber option.

Political reasons

The danger to the public was sited as a reason for cancellation. Accident experiences from other development and production programs was considered. Crashes would result in a major concern with the public. Then, the danger of radioactive contamination from a crash might isolate areas of the countryside (or a city) for years to come. A physicist stated that he did not think that the program can be made to work safely at any reasonable cost. The proper selection of bases and flight routes was considered to offset these concerns. But one advantage of the ANP that was a major driver was the ability to roam the world, without concern about distances or routes. Allocating only certain routes that the nuclear- powered aircraft could fly would defeat one of the drivers for the program. Another problem involved the crew and ground support personnel. One suggestion was that only older pilots and ground crews be employed to minimize the radiation concern. Older people are considered more resistant to radiation than younger people are. Would these people be exposed to radiation levels higher than was safe for them? The concern for future lawsuits was a factor here. Another concern was the possibility of terrorists focusing on a nuclear powered aircraft. Bringing down a nuclear powered aircraft near a city would be a disaster. Prior to President Kennedy coming into office in 1961, Dr. York and his staff again reviewed the program. Doctor York had previously been the director of the Livermore Nuclear Research Laboratory, and was well versed with aspects of nuclear technology. The decision was negative. He reviewed this finding with the incoming Kennedy administration. Their opinions were more negative than his. And so, the program was canceled. Low level work did continue for a time on the P & W indirect cycle reactor / jet engine.

According to Dr. York, the history of the Aircraft Nuclear Program provides a classic illustration of some of the forces that drive the arms race. Congress was controlled by the Democrats and the White House by the Republicans. Pressure to put a plane in flight as soon as possible also doomed the program. The part of the program to develop reactor materials capable of reliable operation over years of operation was not yet successful.

The scientists and engineers working on this project did make significant progress on the technology of nuclear power plants for flight. The failure for this program cannot rest with the technologists.

The engineers who worked on the nuclear program at GE and elsewhere, moved to other high technology programs: the nuclear- powered ramjet and rocket programs that were still being developed; nuclear submarine construction; and commercial electric generating power plants. The AEC explained why ANP had been successful in advancing technology. "the Aircraft Nuclear Program was starting at the upper limits of nuclear technology. These required many so-called breakthroughs in materials, reactor concepts, instrumentation, shielding, and controls. These improvements provided a tremendous acceleration in the advancement of nuclear reactor technology." The people who worked at GE-ANP Department were highly trained, disciplined, and eager to produce excellent work in other technology fields. (i.e. Space/Apollo Programs)

Moving the HTRE-2 to its final resting place for display in 1985. Arco Idaho

INEEL Photo

The article below appeared in Current Science Vol. XLIV, NO 22 March 1959.
What is silly about the drawing is that it shows the A/C carrying a spare engine!

What About the A-Plane?

Much secrecy surrounds America's A-plane project. Here's what CURRENT SCIENCE has found out.

HOW soon will the United States have an atomic-powered aircraft? Why has it taken so long to develop such a plane? What sort of research is in progress?

An atomic plane has many advantages over ordinary aircraft. It can fly almost indefinitely without refueling. An ordinary plane might land on its back and furnish it with food, supplies, and fresh crew members. In the event of a surprise attack, enemy missiles would not be able to destroy an airborne squadron of A-planes.

But the reporter who tries to learn new information about the development of atomic-powered aircraft may become discouraged. Progress that has been made is kept secret. He hears once again that there are basic problems to be solved before nuclear propulsion for aircraft is a reality.

Last December, the magazine *Aviation Week* created a flurry of excitement by reporting that Russia is now flying a nuclear-powered airplane. Our country has flown a conventional plane, the B-36, with a reactor aboard, but it has had no flights by a plane powered with a nuclear engine. The *Aviation Week* story caused Americans to ask questions about their own progress.

The Problems

Three major problems face the men whose job it is to design a workable and practical nuclear-powered aircraft. Few difficulties are concerned with the engine itself. The problems are all directly related to the structure of the entire plane, or the air frame.

Reactors are large and clumsy. The size and weight of a reactor are great. This bulk must be reduced, or else an air frame must be devised that is capable of supporting the heavy power plant. In addition, the air frame must carry a payload of passengers, weapons, or cargo.

Anti-radiation shielding is heavy, but essential. The plane's crew must be adequately protected from the harmful rays given off by the reactor. The air frame itself, as well as the crew, could suffer from extensive exposure to radiation.

Terrific heat is generated by nuclear fuel. Satisfactory materials to withstand long hours of operation must be found. Otherwise, the power plant must be operated at low temperatures which reduce its efficiency.

Last fall, there were indications that at least one of these three basic problems had been solved by the Battelle Memorial Institute of Columbus, Ohio. Experiments there are being made for Pratt & Whitney, a major manufacturer of aircraft power plants. Battelle's researchers are working on a classified project; they do not wish to talk about the progress they have made. However, they were willing to discuss with your CURRENT SCIENCE AND AVIATION reporter another project which concerns atomic-powered flight.

Battelle scientists have begun research to determine how nuclear radiation may affect the "bottled" oxygen supply of aircraft crews. The one-year study is being sponsored by the Air Force's Air Research and Development Command. It is to learn whether oxygen used by crews for breathing at high altitudes might be made unusable when stored outside the shielded area used for crew members.

Once the "building" problems are solved, the "flying" problems promise to be almost as complex. The crew members will be adequately protected as long as all goes well. But suppose the plane should crash? The release of fission particles might endanger the public.

Major General Donald J. Keirn, chief of the nuclear-plane project, says that the Air Force plans to apply very strict flight rules to nuclear-powered aircraft. The planes will be flown only at certain altitudes over specified airways or corridors. Their landing and take-off will probably be restricted to special airfields.

The Second Dozen Years?

In 1946, the first United States project for studying atomic planes, NEPA (Nuclear Energy Propulsion for Aircraft), was organized. In the years since, $700,000,000 has been spent on various phases of the project.

Some experts feel that the A-plane could be just around the corner. Colonel Leonard F. Harman, applications engineer with General Electric's aircraft nuclear propulsion department, recently gave his views. He said, "There is no technical reason why suitable nuclear-powered aircraft cannot be immediately reduced to practice. . . ."

Nevertheless, questions still remain. Will there be an atomic plane in the next few years? Will sufficient funds be appropriated for a crash program? Will another dozen years be spent in research projects? Will some better means of propulsion be discovered before the A-plane is perfected? The reporter who tries to learn the answers is apt to become baffled.

DESIGN FOR ATOM-POWERED BOMBER. ***Building an air frame for a nuclear-powered plane raises many problems. This diagram is an artist's conception of the interior of such a plane. Propulsion takes place as follows: Fissioning uranium in reactor produces great heat; liquid carries heat to heat exchanger; liquid carries heat to jet chamber; hot air expands to provide jet exhaust.***

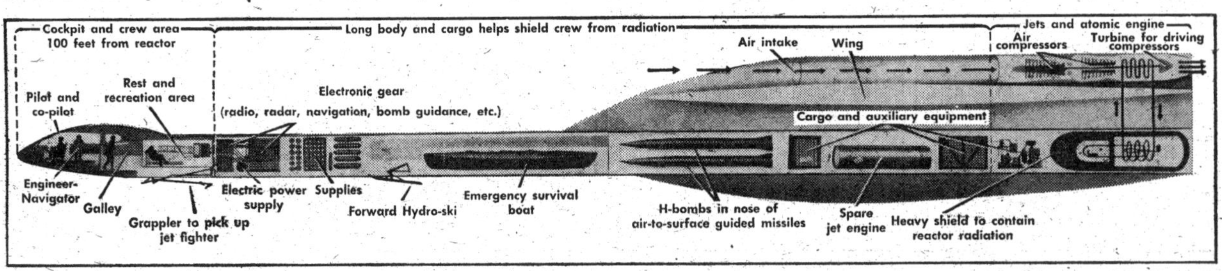

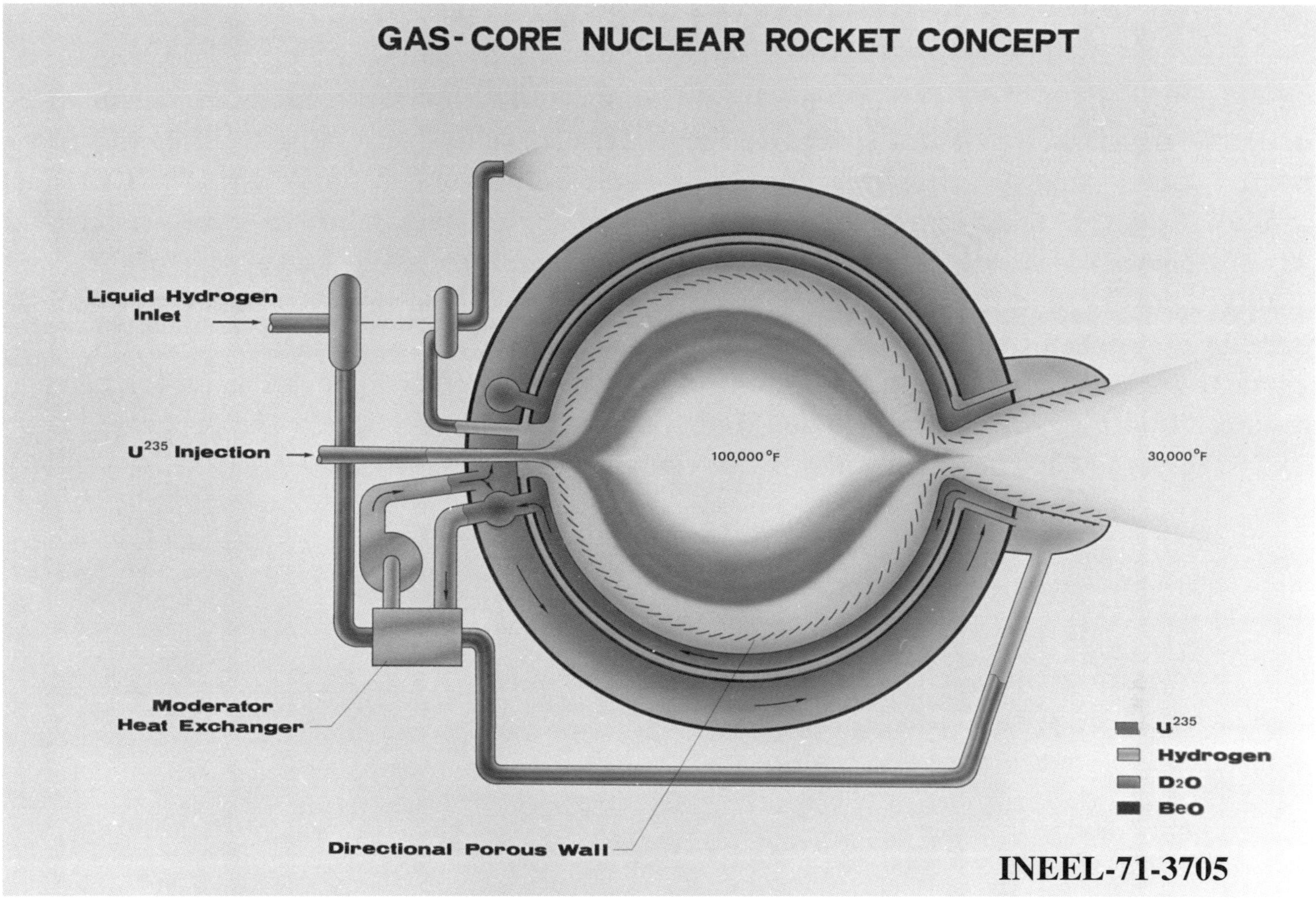

Above. Gas-core nuclear rocket concept. Uranium-235 gas is in the center like a bubble. Hydrogen flows around the uranium, heats up, and exits the core through a small opening, providing thrust.

Unique follow on work at the Idaho test site included the above concept. To quote Jay Konze a GE ANP physicist and engineer.

Eventually, we got into space electric power and worked on a thermionic reactor, in which the fuel elements were thermionic cells. The fuel gave off electrons to supply heat. Then Phillips experimented with the "710," a high-temperature reactor for rocket propulsion. The concept never went into operation.

Later, we investigated another concept for NASA rocket propulsion. This was for a manned mission to Mars. NASA hoped that a one-year mission to Mars using chemical fuel could be reduced to three months on nuclear fuel. it was a cavity reactor, a sphere about twelve inches in diameter. In the center was the fuel, uranium hexafluoride, which above room temperature, is a gas. Hydrogen would flow around the chain reaction in the fuel, heat up to temperatures up to 20,000 degrees F. and exit through a small nozzle, providing thrust.

Thus, some replacement research came to Idaho. The new work made some use of TAN's empire of buildings. The hangar had never been used. The government had poured over $41 million into the Idaho ANP buildings and facilities through 1961. NASA put on hold its plans for a manned mission to Mars, so the Cavity Reactor and the other space-related reactors were shut down in the early 1970s. The vacant TAN facilities went up for rent, a testimonial that the NRTS, no matter how brilliant its scientists and engineers, could not control its destiny when the political winds of Washington blew across the desert.

ACKNOWLEDGEMENTS

Special thanks to George Pomeroy, John Collins, and Otto Woike, members of the original ANP program for their assistance as reviewers as well as providing tech/photo support. A thanks goes to Jim Stump for his fine editorial review. John Walsh of INEEL out in Idaho was very supportive with text and photos. Brian Nicklas, at the Smithsonian Air and Space Museum, came through as always with his excellent knowledge of the archives library, and leads to other sources. Sandy Moltz was very key in early research back in 1997 when the book was first started. Col. Sigmund Alexander, USAF (RET) was an early supporter. Thanks to David Simson of the New England Air Museum for his research findings. Al Liptak supplied some interesting input early in the program. And lastly thanks to Corrine Johnson for her early nod of support.

GLOSSARY

AEC	**Atomic Energy Commission**
ANP	**Aircraft Nuclear Propulsion**
ANPD	**GE's Aircraft Nuclear Propulsion Department**
CANEL	**Connecticut Aircraft Nuclear Engine Laboratory**
DoD	**US Defense Department**
GAO	**Government Accounting Office**
HTRE	**Heat Transfer Reactor Experiment**
ICBM	**Intercontinental Ballistic Missile**
NACA	**National Advisory Committee for Aeronautics**
NEPA	**Nuclear Energy for the Propulsion of Aircraft**
NRTS	**National Reactor Test Station (Idaho)**
INEEL	**Idaho National Engineering & Environmental Laboratories, (Proving the Principle) 1995**

APPENDIX

92A

NUCLEAR POWERED AIRCRAFT

By

J. F. BRADY

Convair, Division
General Dynamics Corp.

For presentation at the
SAE NATIONAL AERONAUTIC MEETING
The Ambassador, Los Angeles, California
September 29-October 4, 1958

Written discussion of this paper will be accepted by SAE until Oct. 31, 1958. Three double-spaced copies are appreciated.

SOCIETY of AUTOMOTIVE ENGINEERS, Inc., 485 Lexington Avenue, New York 17, N. Y.

NUCLEAR POWERED AIRCRAFT

Opening Remarks

The phrase "Nuclear Powered Aircraft" has changed over the years, from a phrase that sounded intriguing, intensely interesting, and excitingly romantic - to a phrase ringing of stupid politics, blood and guts engineering, and just plain old hard work.

I have had the pleasure of being a part of the "Nuclear Powered Aircraft" Program for the past 5-1/2 years, with all of its ups and downs - and there have been many downs.

I have watched, and had a part in, the hiring into the large aircraft companies, of the young brilliant nuclear physicist, with the fate of the world resting squarely on his intelligent forehead, only to watch him grow unhappy, disgruntled, and finally quit because of the slow progress, or many program cut-backs.

Today, the program is limping along, licking its wounds, and making slow but steady progress. In this presentation I will give, within the limits of security, some data on the different types of problems encountered and some idea of the types of aircraft being considered.

We will first take a look at schematics of the proposed power plants. Fig. 1 shows an over-simplified schematic diagram of a reactor. It is made up of a central core where the fuel (uranium) is located, and where the heat producing fission process takes place. Heat is removed from this core by a coolant such as air, water or liquid metal. This core is surrounded by a moderator, whose basic function is to moderate (slow down) and reflect the neutrons back into the core at the correct velocity to continue the fission process. Because there is a large leakage of neutrons and gamma rays from the core through the moderator, it is necessary to shield the reactor and moderator. The shielding of gamma rays requires a very dense material, like lead, and therefore it is located close in to save weight. The neutron shielding is lighter (water as an example) and is located farther out.

Fig. 2 shows some of the proposed propulsion cycles for aircraft. In the upper left corner we see the most direct application: that of heating the propulsive air by directing it through the reactor. Air enters the jet engine, is compressed by a compressor in the usual manner, then passes through the reactor where the heat of fission is transferred directly to the air; then expanded through a normal turbine and expelled through the tail pipe producing thrust. This jet engine can have the usual afterburner burning ordinary aviation fuels, if desired.

The turboprop version of this cycle is shown in the lower left hand corner, the only difference being the transfer of power through a gear box and thence to the propeller.

Shielding is difficult for this cycle because of the large air passages required through the shielding.

On the right hand side of the illustration we have another possible cycle, where the reactor heat is transferred to a medium which in turn transfers its heat to the engine air (similar to the radiator in your car). One of the advantages is the use of a dense heat transfer medium resulting in smaller reactor heat transfer surfaces and therefore smaller reactor, and also smaller holes through the shielding. A disadvantage is the plumbing involved and the loss of heat because of two transfers.

Either of these cycles can use normal aviation fuel in addition to or in place of nuclear power as shown in the central figure.

The next few illustrations will indicate some of the problems associated with designing nuclear aircraft. The problem that designers and physicists have been battling for years is the shielding problem. As previously discussed we indicated the necessity for heavy lead and water shielding to capture the neutron and gamma ray radiation. Fig. 3 shows two methods of shielding: on the right the unit shield where all the shielding is at the reactor, resulting in low radiation throughout the aircraft as indicated by the light color. On the left we see the divided shield concept where the shielding is distributed around the reactor and also around the crew. This latter method is lighter (provided the crew is small enough - usually the case in combat type aircraft); however, it has the disadvantage of allowing a higher dose to escape from the reactor and therefore the aircraft structure and components are in a higher radiation field. Also this makes it difficult to handle the airplane.

The curve shows the divided shield being lighter and also the trend of getting more power per unit of shield weight as the power is increased. This, of course, tends toward larger airplanes, but other factors, such as handling, cost, etc., prevent us from taking full advantage of this.

Radiation of the aircraft components and structure also causes damage. Fig. 4 shows some of the more common materials used in airframes and components and their approximate life under normal estimated radiation fields associated with the divided shield type aircraft. It is obvious that no structural problem exists; however, oils, lubricants and rubbers will require some development. Electronic systems are not shown because of the security involved, but many problems exist; however, programs are underway to solve these problems.

Another problem to contend with is the aircraft activation. Neutron bombardment of the airframe components and structure causes them to become radioactive. When the activating source is removed the activated part continues to radiate but there is a reduction of radiation per time. This is indicated by Fig. 5. This illustration shows typical airframe activation per time after reactor shut-down, and with the power plant removed from the airframe. The arrows indicate areas outside of which a man can work under normal laboratory tolerances as set up by AEC. This does not mean there is any danger inside these areas; it only means the men would work a shorter time than the usual 40 hour week and reasonable personnel rotation could handle this problem. It must be remembered that this would vary greatly depending upon the amount of shielding used. A unit shielded airplane could be designed to have no such activation problem.

Having discussed the nuclear power plants and the major nuclear problems in broad general terms, now let us be a little more specific and see what types of aircraft might result. Fig. 6 shows a typical attack type aircraft. It is a delta wing airplane utilizing a direct air cycle type power plant as shown schematically in Fig. 2. This type aircraft requires a highly divided shield resulting in high dose rates throughout the aircraft; however, the crew is housed in the crew shield and is subjected to reasonable dose rates. These dose rates are classified; however, they can be given in allowable pilot nuclear cruise hours per year as indicated. The next illustration will discuss this in more detail.

This airplane is designed to either penetrate at high subsonic speed at sea level or supersonic speed at high altitude. The low altitude mission can be performed on nuclear power only; however, the high altitude supersonic mission requires chemical fuel augmentation. This is reflected in the increased weight when this fuel is added. This brings us to one of the undesirable present day limitations of nuclear power - that is, maximum operating temperature. At present, and for some years to come, reactors have a maximum operating temperature resulting from metallurgical limitations. The turbo jet engines being coupled with the reactors do not have this temperature restriction, and can therefore use chemical fuel to develop higher thrust when desired.

The radius of 8000 n.mi. and 500 n.mi. - as shown - are for broad general range indications only, and could be easily extended with a small increase in dose rate to the crew.

Fig. 7 indicates the dose to the pilot of the aircraft described on the last illustration. After looking at the career of the average naval aviator we concluded that with this dose limit it is reasonable to stay within National Academy of Sciences and National Bureau of Standards radiation limits. The lower of the two (National Academy of Sciences) is indicated on this illustration. By rotating the pilot's duty between shore duty and active nuclear aircraft operation, total dose for his career will be 90 REM. This leaves at least 10 REM margin for background radiation, X-Rays, etc. It should be noted that the system of duty rotation is not unusual. It is done as a normal procedure.

The number of hours of nuclear flight per year that the pilot would be permitted to fly, with and without augmented shields, is shown in the lower right hand corner of the illustration. The augmented shield would be increased in weight by the amount of payload, for pilot training on nuclear power. These numbers are considered reasonable in light of the fact that present day non-nuclear aircraft of this category rarely exceed 300 hours per year.

The nuclear turbojet powered attack aircraft uses a divided shield - part of the shield material is on the reactor and part around the crew compartment - in order to avoid excessive weight penalties. As a result, the radiation dose rates are quite high outside of the crew compartment during reactor operation.

Fig. 8 shows the estimated service life of various materials at several locations in the hull. Many materials offer no problem, of course, but others may either require further development to increase their resistance to radiation or special effort may be required by the designer to see that the sensitive materials are not used in high dose areas.

You will recall that one of the problems associated with nuclear propulsion is activation and after-shutdown radiation. For the attack aircraft using the direct air cycle turbojet considerable gamma radiation results from fission product decay in the reactor after shut-down. The table on Fig. 9 shows the after-shutdown activity from the airframe only in millirem/hr. Laboratory tolerance is presently established at 300 millirem/week. The power plant has been removed from this aircraft in order to perform service and maintenance within practical time limits. With the engine installed this is impossible. In actual practice some functions such as servicing for rapid turnaround would probably be done by remote procedures while other functions such as periodic inspections would be done with the power plant removed.

Fig. 10 shows a nuclear turboprop powered airplane which might be used for a number of missions such as ASW, AEW, barrier patrol, or stand-off attack. This aircraft has a unit shielded direct air cycle reactor made possible by separating the reactor from the engines. The dose to the crew is low enough that the endurance of the aircraft is limited only by the physical endurance of the crew. Also, there would be no radiation activation or damage resulting from such a low dose.

This aircraft weighs greater than 400,000 lb., but it should be emphasized that it has not been optimized to an optimum power plant. Instead it was designed to use what is considered the earliest nuclear power plant which could be made available. The aircraft could have been considerably smaller with an optimum power plant.

This power plant consists of (4) existing turboprop engines modified for nuclear power by removing the burners and installing scrolls to pipe the engine air from the compressor to the reactor and back to the turbine. The reactor itself is a low temperature adaptation of current state-of-the-art development of the direct air cycle.

This aircraft can also take advantage of reactor development advances and higher temperatures without becoming obsolete. By using higher temperatures, smaller reactors can be used thus increasing the payload.

In summary, two possible programs can be pursued toward development of nuclear powered aircraft. They are:

Turboprop

1. The fastest, easiest program would be a subsonic turboprop aircraft because it would use existing components and technology.

2. Low radiation from the unit shielded reactor would eliminate radiation activation and damage thus minimizing the operating and handling problems.

3. It is also important to note that this would be a logical development program. Advanced data on powerplant operating techniques, servicing and maintenance problems and general operating problems associated with a nuclear reactor can be investigated at low risk to personnel and equipment.

Turbojet

1. The turbojet powerplant is now under development but needs an advance in technology over the turboprop powerplant.

2. The operating and handling procedures necessary to obtain optimum use of the turbojet aircraft need investigation and development which can be accomplished in part with the turboprop aircraft.

If only our politicians, military leaders, and numerous Department of Defense committees would realize that even with our first crude power plants we can show useful nuclear powered aircraft - and if they would only remember the utterly useless Wright Bros. airplane - and if they would only remember the low performance of our first jet powered aircraft - and if they would realize that these embryo beginnings are developing a knowledge of almost unlimited possibilities - then maybe they would get off their broad backsides and help this country to be first with the Nuclear Powered Aircraft.

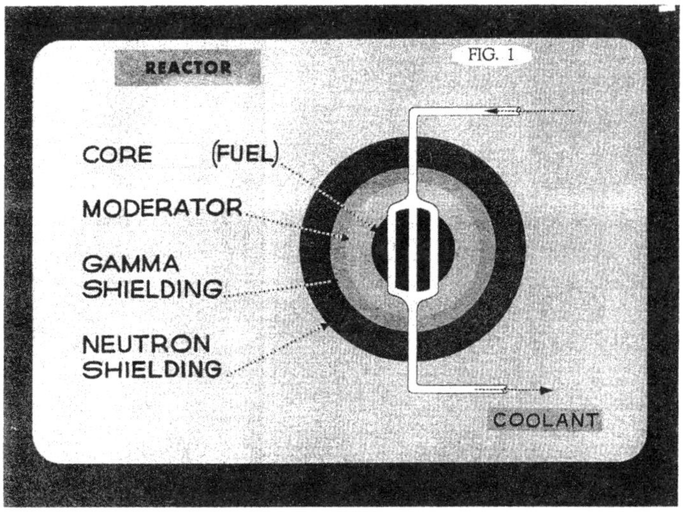

FIG. 1

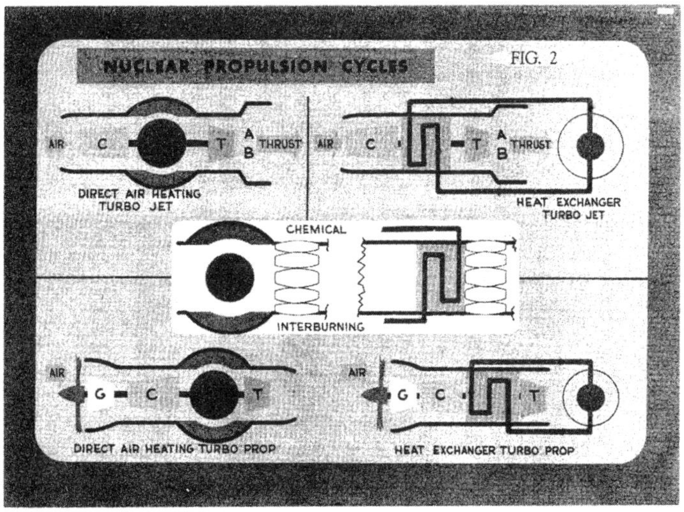

FIG. 2

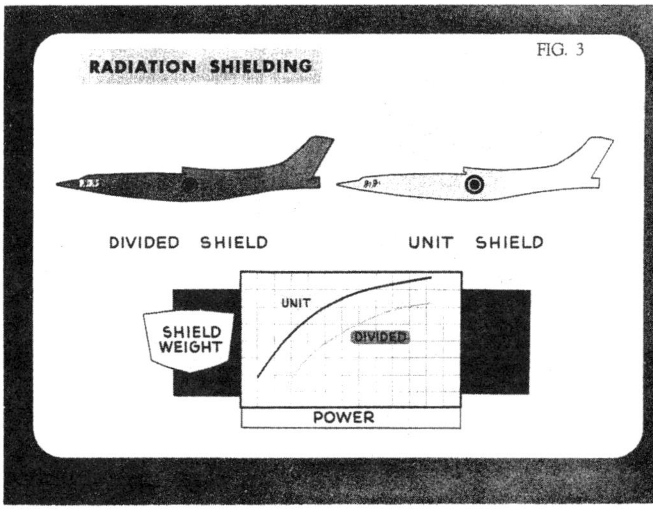

FIG. 3

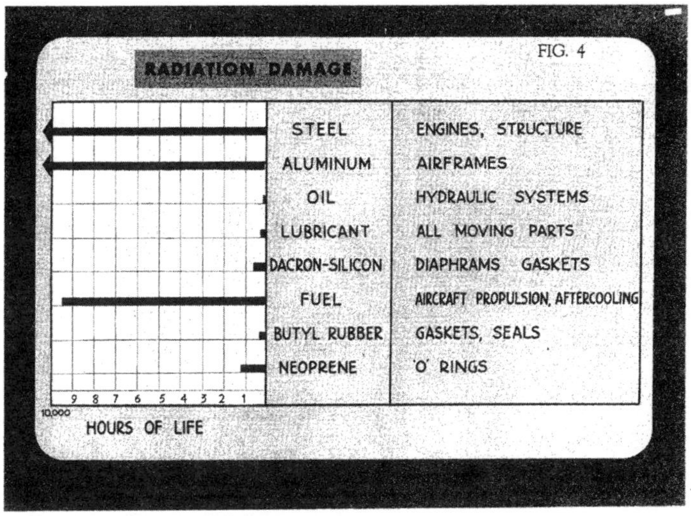

FIG. 4

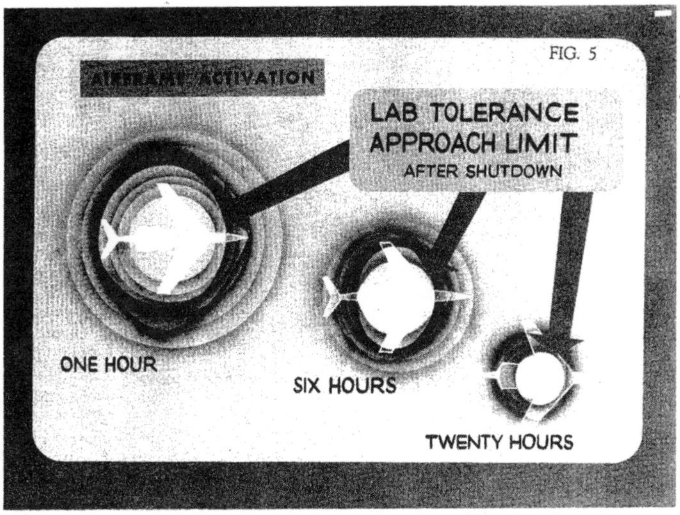

FIG. 5

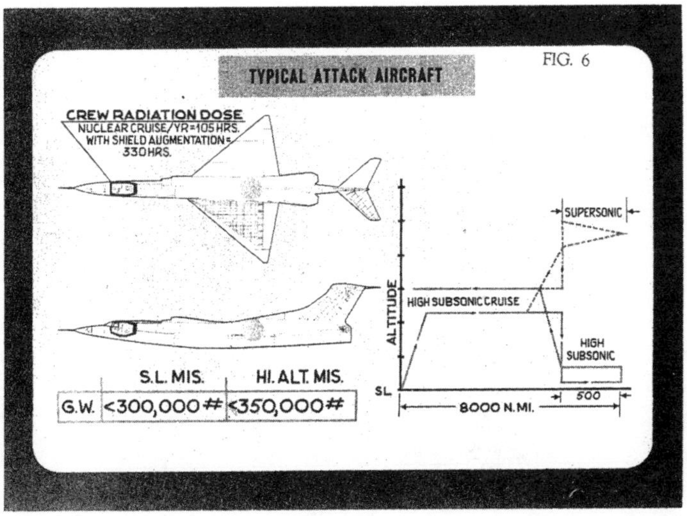

FIG. 6

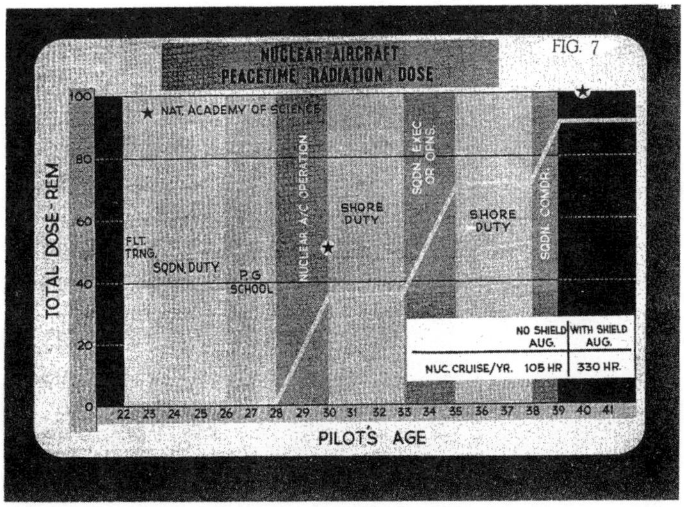

FIG. 7

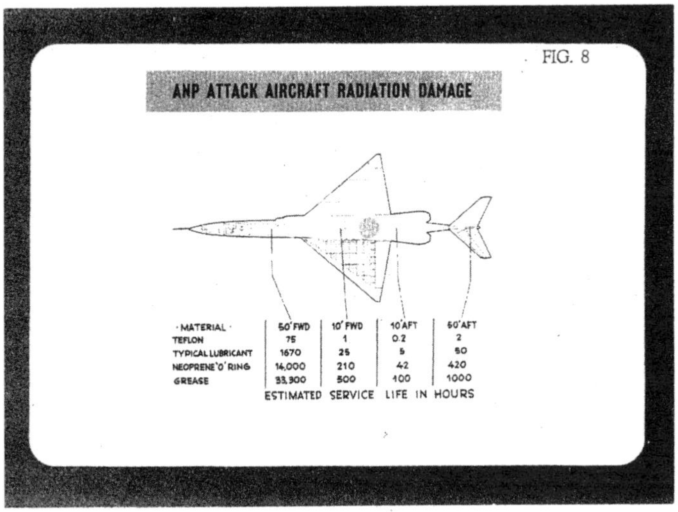

FIG. 8

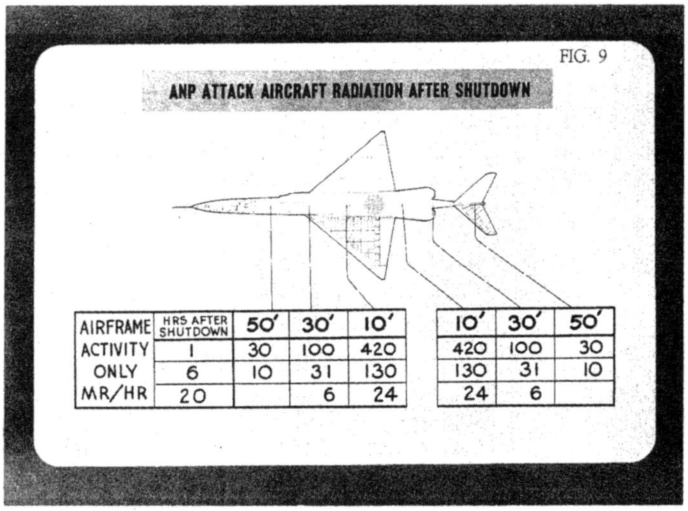

FIG. 9

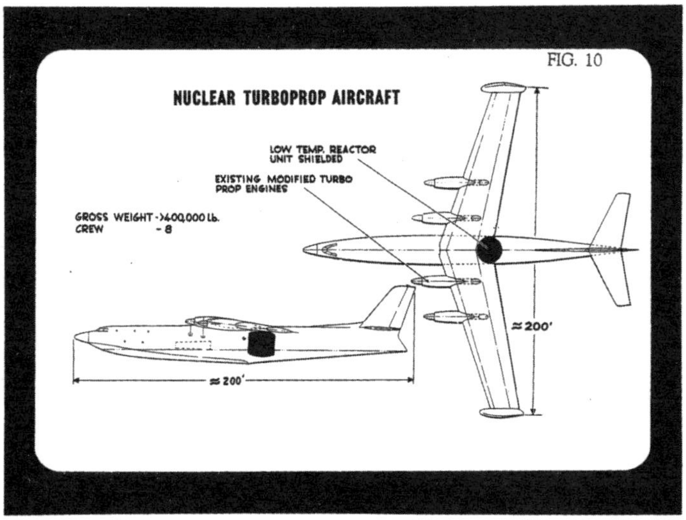

FIG. 10

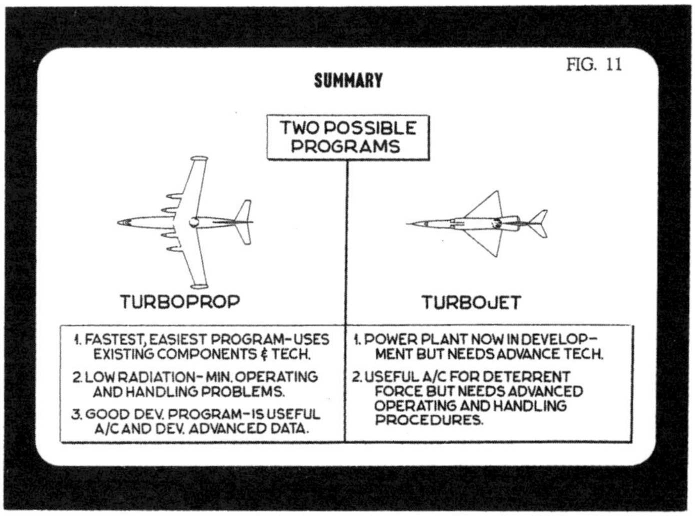

FIG. 11

APPENDIX

The following is a speech from 1949 given by Lawrence R. Hafstad, Director of the AEC.

Very Insightful Author

UNITED STATES
ATOMIC ENERGY COMMISSION
Washington 25, D. C.

INFORMATION FOR THE PRESS
Telephone ST 8000, Ext. 307

FOR RELEASE A.M. PAPERS
Saturday, July 23, 1949

REMARKS BY LAWRENCE R. HAFSTAD,
DIRECTOR OF REACTOR DEVELOPMENT
UNITED STATES ATOMIC ENERGY COMMISSION,
BEFORE THE INSTITUTE OF AERONAUTICAL SCIENCES
LOS ANGELES, CALIFORNIA, JULY 22, 1949
(Ambassador Hotel--7:30 P.M. PDT)

ATOMIC POWER AND ITS IMPLICATIONS FOR AIRCRAFT PROPULSION

It is indeed a privilege to meet with this particular group tonight to present my impressions of the problems involved in the development of nuclear reactors, first in the general sense, and second, for the specific purpose of use in aircraft.

This group is especially qualified to appreciate the nature of the problems involved, both from a technical and from a policy point of view. Reactor projects have much in common with aircraft development projects. Both are expensive and risky. In both fields, one must obey the precept "Be bold, be bold and ever more be bold, but be not too bold". Following that advice takes more than technical knowledge. It requires wisdom.

This report tonight is timely, too, for since I have been with the Commission just long enough to learn the ropes, I should be able to give you the impressions of an outsider who has had the privilege of getting a good, hard inside look at the Commission's Reactor Program.

Of course, this assumes that I will be skillful enough to handle the security questions involved. Fortunately, I do not think this will be too difficult. What our competitor wants to know is, "What solutions or promising approaches have we found to specific technical problems?" What you gentlemen are probably most interested in is: "Is the objective we have set for ourselves worth attaining and are we going about it the right way?" These are questions peculiar to our own situation and our own frame of reference. The answers we get are certainly not those which are applicable to our competitor. In this area, therefore, we can afford to be frank without giving aid and comfort to the competitor.

(more)

Let us see if we can get squarely in mind just what are we trying to do in the overall atomic energy effort? Not with respect to specific technical projects, but what are the things we have to do in a broad sense? These are, I would say:

First: To maintain undisputed leadership in technical knowledge and facilities for the United States in all areas of the atomic energy field, either for war or peace.

Second: To establish a trained corps of personnel familiar with all areas of atomic research and in particular with nuclear radiation and its hazards. This corps is our standing army of the modern age, and would have to be maintained to serve in time of war, even if not organized to contribute in time of peace.

Third: We want organized task forces sufficiently competent to explore, and sufficiently alert to exploit any technical advance, which has promise for either peace or war.

Fourth: We want the wisdom to see that the effort expended in the above activities is somehow kept proportional to the probable returns.

As part of the above, some kind of a reactor program is an essential part of the overall effort. Reactors are the machines for giving us the controlled release of nuclear energy, and as such, certainly have promise both for peace and for war. Though the spectacular aspects of nuclear energy has been oversold in the popular press, the stubborn fact remains that one pound of Uranium can be persuaded to release an amount of energy equivalent to 2,000,000 pounds of coal. When all is said and done, atomic energy is certainly not the magic perpetual motion machine which has been publicized, but it has the inherent possibilities of providing an incredibly compact storage battery.

Let's look at this analogy a little closer. Our ordinary automobile battery delivers 6 volts and has a capacity of about 100 ampere hours, or a total power storage therefore of very roughly, one horsepower hour. A weight of Uranium equal to that of an automobile battery would be capable of delivering about 300,000,000 horsepower hours. Edison was a great inventor, but the famous Edison cell, produced after years of effort brought forth an improvement of only 30 percent over the conventional lead cell compared to a conceivable 300,000,000 percent

(more)

for the Uranium energy source. I will return to the many difficulties later, but this analogy shows the challenge of the problem, the conceivable rewards, the "pie in the sky." It would seem that the possibility of nuclear reactors for power production must at least be explored. The '49ers had much less incentive.

What else can reactors be used for? In addition to energy, reactors produce neutrons. These in turn can be used to produce radioactive isotopes, or to produce more, or other, fissionable material. So long as concentrated energy sources are desirable, fissionable material will be an "economic good." A stockpile of fissionable material would certainly be more useful than the gold at Fort Knox. In war, it could be used for bombs and might be used for propulsive power. In peace, it would be available for civilian power insofar as the supply, and the economics of the situation would permit. It would appear, therefore, that a stockpile of fissionable material is indeed desirable. Accepting this, then considerable effort would appear to be justified on a special type of reactor called a "breeder," which shows promise of helping to maintain, if not even augmenting, our supply of fissionable material.

To describe a breeder we must digress for just a moment to a technical detail. As most of you know, the element Uranium, as it occurs in nature, is composed of two kinds, one variety slightly heavier than the other. The fissionable variety from which we can get energy has a mass of 235 units. The other, unfissionable, has a mass of 238 units. The U-238 form is 139 times as prevalent as the U-235 form. The trick in a "breeder" type of reactor is this: With a proper choice and arrangement of materials, the neutrons which are produced in the initial natural U-235 fission processes can be captured--<u>after</u> they have released their energy--by the atoms of unfissionable U-238 and as a <u>by-product</u> used to convert the U-238 into a fissionable form of material. Thus it is conceivable, not only that energy can be extracted from U-235, but that as a by-product we acquire a potential stockpile of fissionable material 139 times as great as we had when we started! With such a stockpile we might have a chance for diversions to power for civilian uses. Again we have a challenging goal and unless we have lost our zest for adventure, it is a goal from which we will be deterred only by a convincing demonstration of its scientific impossibility.

If reactors are so desirable, why don't we go ahead and build some? Now we come to the difficulties! Here are a few of them:

(more)

1) That for any reasonable thermodynamic efficiency in utilizing the great energies available, it is necessary to operate at temperatures well above the conventional engineering range.

2) The compactness of reactors which is an important inherent advantage proves troublesome in regard to the heat transfer problems which involve heat transfer rates far transcending previous experience.

3) The materials chosen for the reactor must withstand not only high temperatures but also high nuclear radiation densities, with unpredictable changes in the physical properties of the materials concerned. The seriousness of this problem can perhaps be visualized by this kind of a comparison. How would you like, for example, to design airplanes or engines if, in use, the properties of the aluminum and steel would gradually change to those of cast iron or lead?

4) If we finally find a structural material for reactors which appears suitable so far as physical properties are concerned, we must now add still another requirement. The nuclear properties must be such that the structural material will not capture neutrons and thus deplete the supply and reduce the power. This requirement drives us to consider strange new elements, and raises a whole array of procurement problems.

5) Even after we have our reactor working we find that the fission products produced as an essential part of the reaction "poison" the reaction itself. The ashes smother the fire. Now you gentlemen are well aware of the enormous maintenance problems for aircraft engines. Every 800 hours they must be disassembled, inspected, have defective parts replaced, and then reassembled and tested. The work is staggering. However, how would you like it, if instead of merely disassembling, the entire engine would have to be dissolved in nitric acid, and the rebuilding of the engine started with getting a solution of certified chemically pure iron? This is the fuel reprocessing problem!

6) Finally, assuming we have solved the structural problems listed above, we have a whole new category of problems in connection with the working fluid or heat transfer medium used to convert the heat into power. The nature of these problems can be suggested by the fact that from rough comparisons of the volumes of

(more)

reactors and the present highly perfected aircraft engines, the rates of heat transfer must be more than an order of magnitude greater for nuclear reactors than for conventional engines. Orthodox advances will not be sufficient. The problems involve the use of liquid metals with all the associated corrosion, erosion, purification and pumping troubles which we can readily imagine as being associated with those elements which appear to have suitably low melting points.

When one considers the host of difficulties and troubles which lie in the road ahead in the development of atomic power the problem does look formidable. I am reminded of a statement made a little over a hundred years ago by the great chemist, Wohler, in regard to the status of organic chemistry at that time. Wohler wrote to Berzelius as follows:

> "Organic chemistry just now is enough to drive one mad. It gives me the impression of a primeval tropical forest, full of the most remarkable things, a monstrous and boundless thicket, with no way of escape, into which one may well dread to enter."

That's an excellent description of the atomic energy field right now, in 1949! In the meantime, however, what has happened to organic chemistry? Well, newspaper headlines give the answer. Miracle drugs are practically tailor-made these days. DDT and 2-4-D are taken for granted by the farmer. Synthetic rubber threatens to displace the natural product. A hundred years from now what will be the status of atomic energy? Who now has the wisdom to predict either failure or success?

We can all hope for the era of free power and effortless living usually associated with the Atomic Age. This implies the successful development of large, land based, electric power producing reactors. We have also heard discussed the military advantages which might be gained by nuclear propulsion of ships and aircraft. I will discuss these in more detail later but the point I want to make now is that whereas the technical problems would be least in the land based power reactor, and progressively more difficult in the ship and aircraft reactors, the present urgencies or priorities are just the other way around. Perhaps fortunately, however, the same ground must be covered in the initial stages whether the ultimate purpose is for civilian or military use. We might take as an analogy, a transcontinental journey, starting from Washington in the frontier days. Whether the ultimate goal was Oregon or California, the route was the same through Cumberland Gap and on

(more)

to St. Louis. To complete our comparison we might put civilian power in California and military power in Oregon. These are the things we dream about. At the moment we are really only approaching Hagerstown, and our worries and our plans are all concerned with surviving the hazards of the journey to St. Louis.

I don't need to stress before this group the importance of the incentive given to technical developments by military needs. This group is well aware, I might even say, painfully aware, of the vicissitudes of the development of the airplane to the highly perfected state in which we have it today. Similarly in the atomic power field it appears that military needs will have to provide the incentive to carry through difficulties, for progress even though ultimate dividends may be expected in the civilian economy.

Going further back into history, we can cite the difficulties of converting ships from wood to steel. Again, the incentive was military, but note this quotation:

> "Early experiences with iron as the material for hull construction were far from reassuring. In England where several iron warships had been completed by 1846, firing trials conducted in 1845 and in 1850 indicated that 32-pounder and 68-pounder shot striking iron plating were likely to break up and form more splinters from the shot themselves and the iron of the target than were caused by the impact of the same shot upon wooden targets. Accordingly, the British Navy pronounced iron to be an unfit material for hull construction."

Only if both the opportunities and the difficulties in the field of atomic energy are fully appreciated can the history of the atomic developments over the last several years be understood. This is a field in which the experts disagree. The more distinguished they are the more violently do they disagree. /At this point, I want to disqualify myself as an expert. Scientists live and work in laboratories, not in marble buildings in Washington! I am an ex-scientist now. My job is to be a good listener and an accurate interpreter./ When experts disagree, a middle of the road course of action is indicated and this indeed is what we have in the Commission's Reactor Development Program.

As has been announced, and as presented in Budget hearings before Congress, so there are no security questions involved, the Commission program consists of two main parts. The first is a strong applied research program seeking to establish the

(more)

basic facts, the handbook data if you like, which will ultimately be needed in solving the reactor design problems. For the reactor program the center of this type of authority is at the great Argonne Laboratory at Chicago.

Due to the foresight of the Manhattan District and its advisors, and to the continuing generous support by the Commission and Congress, the nation has an exceptionally well supported predominantly non-military applied research program in the atomic energy field. In the large national laboratories we have thousands of people working on and becoming acquainted with atomic energy problems. In our atomic energy production plants we have thousands more. These people are our standing army, mentioned above as requirement Number 2, and our preparation for any eventuality of the so-called Atomic Age.

But it will take more than the accumulation of a library full of knowledge to get power-producing reactors. We have 40,000,000 automobiles on the roads of the United States but we still do not "understand" the mechanism of combustion. It is for this reason that the second part of the Commission's program is the engineering development and construction of a series of definitely experimental prototype reactors. These represent assignments to specific task forces as mentioned in requirement Number 3.

Of course, the nation has had reactors of various kinds, from almost the beginning of its atomic energy program. The famous chain reacting pile at Chicago was the first of such reactors. A series of reactors was built during the war culminating in the huge single purpose production reactors at Hanford. Other reactors have since been built but these are by and large, research type of reactors, small in size, and none of them capable of producing useful power in appreciable amounts. The next phase in the historic development of reactors calls for designing and constructing of reactors which are larger, more complicated, and more difficult to build than any we have produced thus far. As in any new technical development, there are many uncertainties and many risks involved. It is here that the experts disagree on details of designs of reactors which will do the jobs that need to be done. Largely, for this reason, in the four years since the end of the war, no really new or greatly improved versions of reactors have been built in this country. The reactor of most advanced design and performance is in Canada.

The proposed reactor development program of the Commission crystallized out of the four years of discussion and argument as well as from new knowledge gained from the applied research

(more)

program since the end of the war. Reactors can conceivably be used for a wide variety of purposes. Special reactors of many types have been proposed by responsible people for purposes varying from small compact units for propelling guided missiles, to huge stationary power plants for providing cheap electric power for supplying our great cities and distilling ocean water for irrigating our deserts. To the people most fully informed, it is clear that the difficulties of building any reactor are so great that only a very few projects can be adequately supported with money and particularly with competent technical manpower at the present time. It is for this reason that it is essential from the multitude of possible reactors, only a few carefully selected projects should be chosen and very strong technical support should be focused on these few.

Getting back to the fundamentals, a reactor can be made to produce two things: First, a large number of neutrons, and second, a large amount of heat or power. At Hanford, in the production reactors, the neutron supply is utilized for the conversion of the non-fissionable uranium-238 into fissionable plutonium for use in atom bombs. In the existing Oak Ridge reactor, again the neutrons are used for the production of isotopes for peacetime research purposes. In both cases, the heat generated is wasted--it is lost in water coolant at Hanford, in air coolant at Oak Ridge. At the present time, there are no reactors in existence so designed that the heat produced can be made to serve useful purposes.

An obvious forward step would be the design of a reactor in which the neutrons produce fissionable materials as in the existing production reactors, but in addition, the heat generated is put to work. Unfortunately, scientists and engineers at present do not have enough basic knowledge to design such obviously desirable reactors. Their first step would appear to be to produce a reactor specifically for the single purpose of generating large amounts of heat at temperature which will permit conversion to power. So extensive is our ignorance, however, that even such a simplified design is forcing us into pioneering activities beyond the present boundaries of human knowledge. Before any reactor can be built with a performance appreciably better than those we now have, a large amount of applied research in very specialized fields is necessary. This is the activity with which our laboratories have been preoccupied for the last four years.

We are now at the stage where if we intend to progress further, it will be necessary to find the courage to build a few reactors, to test what we think we know. The reactors in the Commission's program are essentially experimental prototypes.

(more)

None of them can be described as an "end-item" which will drive an airplane or a ship, or power and light a city. Further generations of reactors will be required before such desirable goals can be attained. It is this fact which sets a time scale of ten to twenty years before useful and economical civilian applications of atomic power can be expected. However, if the ultimate goal is ever to be attained, the first steps must sooner or later be taken, and it is these first steps with which we are concerned in the present reactor program.

As described by Dr. Robert F. Bacher, former Commissioner, and a moving force in reactor development work, the current program consists of four reactors:

a. The first of these has been designated as a materials testing reactor. We call it "MTR." While it is itself an experimental reactor, as its name implies, it is intended also to give information on the behavior of materials in reactors so that larger and more powerful special purpose reactors may ultimately be built. This reactor is of particular interest to the Air Force since it represents the boldest step into the unknown which we now dare to take, moving in the direction of compact, high radiation density reactors which must ultimately be developed if the Air Force needs are to be satisfied.

b. The second reactor is a land based prototype of a reactor for use in propelling naval vessels. It would be a simple, single purpose reactor designed specifically and solely for the purpose of producing large amounts of heat, under conditions which will permit conversion to propulsive power.

c. The third reactor is a single purpose experimental reactor designed specifically to give us information about the breeding process. This reactor at the present time is the most likely to demonstrate the actual breeding of new fissionable material. It is, however, neither a high power reactor nor designed for the purpose of demonstrating appreciable amounts of useful power though an incidental amount of power may be produced by a by-product.

d. The fourth reactor is the more ambitious project--the Knolls Atomic Power Laboratory Reactor. This reactor is intended to produce a really significant amount of electric power. At the same time, it is hoped that this reactor will be able to demonstrate at least partial success in breeding. This reactor is, therefore, a very

(more)

complex device since its design represents a compromise between the demands for power production and for breeding. If successful, however, this reactor would represent a major step forward in the direction of the production of useful power without depleting, and perhaps even increasing our national supply of fissionable material for any purpose.

Whether by accident or by design, this program is a reasonable middle of the road program. It represents a balance between reactors contributing to the solution of military and civilian problems, a balance between reactors which use up fissionable material and reactors which promise to replenish or increase our national supply of fissionable material, a balance between a bold attempt to solve immediate problems by the engineering approach as in the Navy reactor and the intermediate breeder, and the more long-term research approach of gaining more information about the behavior of materials under novel, but controllable conditions as in the materials testing reactor, and the experimental fast neutron breeder.

I would like now to comment on the fourth requirement which I mentioned above and which was concerned with getting value received for money invested. How does one put a money value on any new developments particularly on one of a military nature? What, for example, would have been the value of the "Merrimac" to the South had not the "Monitor" come along? What was the value of the "Monitor" to the North in 1862 dollars? If we want to be modern, what was the value of the "Spitfire" in the Battle of Britain, or what is the current value of an atomic bomb?

We have heard much discussion recently of the cost of producing atom bombs but what are they really worth to us--in dollars? It would be helpful to have at least a rough estimate of the present value of an A-bomb, either in dollars or in equivalent divisions or battleships or air-groups. While this may sound difficult, order of magnitude engineering estimates to keep our thinking straight, are not too hard to make.

We know, for example, that at the end of the war, our daily war expenses were approaching $300 million per day. If the A-bombs shortened the war by even 10 days, the entire $2,500,000,000, cost of the Manhattan effort can be written off, and recorded, as a spectacular success, and a <u>value</u>, as contrasted to <u>cost</u>, of at least $1,500,000,000 each to set on the Hiroshima and Nagasaki bombs.

Since warfare seems to be mainly a competitive destruction, we can get another estimate in an entirely different way. Taking the radius of destruction for a bomb as from one to two miles, the

(more)

area destroyed would be approximately six square miles. In an average city the property value runs perhaps $50,000,000 per square mile. The destruction per bomb, therefore, represents about $300,000,000 and gives a figure of the "advantage" to us, and therefore of the value to us, in this insane competition in destructiveness.

We can approach the problem in another way to get another independent estimate. I have heard that a man by the name of Churchill, who seems to have a reputation in these matters, has stated that, but for the A-bomb, World War III would have been underway. Now for those who know how many bombs we have, it should not be too difficult to pro-rate the annual estimated cost of this war among the bombs on hand.

Finally, let us approach the problem in still another way, using an infantile form of operations research. We have been told that one A-bomb is equivalent in effect to 20,000 tons of TNT. Since big, single bursts "over-kill" at the center, let us cut this by some suitable factor, say 10, for example. Now one of our large bombers can carry a pay load of about ten tons, therefore, to carry 2,000 tons would require 200 planes. This is unassailably accurate arithmetic! Large bombers cost half-a-million dollars to build, but with logistic support, more like $2,000,000 in combat. One A-bomb makes one bomber the equivalent of two hundred. Ergo, one bomb is "worth" $400,000,000.

There is another conclusion which can be drawn from this quick operations research calculation. We should think twice or maybe even three times before permitting too casual diversions of material from the stockpile, even for other conceivable military uses. It follows also that civilian power will remain "pie in the sky" unless by a "break" in the breeder program or some other solution to the raw material problem, a more ample supply of fissionable material can be provided. Finally, it follows that this is indeed a game in which if we are to play at all, we play with "blue chips."

In the reactor field what would the nuclear-powered air equivalent of the "Merrimac" be worth in a future contest against a fleet of chemically-fueled airplanes? What would a nuclear-powered naval vessel be worth in a future engagement if it were--as the "Merrimac" and "Monitor" were--one whole generation ahead of the conventional fleets of the day? What would be the value

(more)

... nation if, as has been suggested, atomic energy could be used to evaporate sea water to make the deserts bloom? These are the really fundamental questions, and I am glad that it is Congress' responsibility and not mine to make decisions on them. It is, however, my responsibility to see that we get value received out of each dollar which is appropriated for the reactor program, and this will require more than anything else that effort be kept commensurate with both priority and promise.

I would like, therefore, in view of its special interest to this group, to return to the aircraft propulsion reactor to consider it more in detail. To this group, fully aware of the serious limitations of chemical fuels, I feel sure that the desirability of an ideal nuclear power plant for aircraft is obvious. I can, however, quote a Congressional report on the subject. In the Brewster report we find the following statement:

> "In the event of war or in any international situation likely to lead to war, nuclear energy for the propulsion of aircraft would be comparable in significance to the atomic bomb itself. Presently known limitations inherent in all chemical fuels make difficult the delivery by air of atomic bombs against a distant enemy. Therefore, if the United States had nuclear energy propulsion in addition to atomic bombs, it would be the dominant factor in maintaining world peace. Until these ends are attained, the United States must depend on military weapons and techniques currently available."

With the desirability of an ideal solution to this problem there is general agreement. There is agreement too in regard to the contention that developing any kind of an aircraft reactor will be extremely difficult. The NEPA Project, carried out by the Fairchild Company, under an Air Force Contract, has been engaged in a vigorous attack on this problem since 1945. The North American Aviation Company and the Rand Project have also made important contributions. These studies all seem to indicate that, granting the difficulties of the reactor problem itself, the power conversion problem represents a challenge of almost equal magnitude.

As seems to be characteristic of this field of activity, anything which is obviously desirable and important seems to be almost incredibly difficult. To help resolve the impasse, the Commission last year made a contract with the Massachusetts Institute of Technology to make a study of the problem and to come up with recommendations. The result of this study was the Lexington Report, the details of which are at present quite properly highly classified.

(more)

The immediate course of action indicated in the Lexington Report is essentially that the aircraft propulsion project should be continued in an intensive study phase, both theoretical and experimental, for the next two or three years, by which time it might be hoped that data might become available to permit a reevaluation and a more decisive conclusion. It is recognized that this study phase should be made national in scope to include the NACA and AEC as well as the National Military Establishment.

These are eminently sound guideposts and they are being followed. An Ad Hoc Committee, consisting of representatives of the Air Force, Bureau of Aeronautics, NACA, and AEC, has been meeting since last January to coordinate the work of the various government agencies involved and to insure an industry wide approach to the technical problems. In such a joint attack on a problem clearly there will be some duplication which must be eliminated and some shifts of emphasis which somehow must be consummated. Committee procedures grind slowly but this work is well under way. Many of the companies represented here tonight have recently contributed technical talent to the National effort now being organized.

In the meantime, in accordance with the Lexington Project recommendations the NME has been asked to evaluate the military worth of the proposed weapon if and when it is produced. This is a really tough assignment. It is indeed controversial, but not in the sense of an inter-service feud. This is definitely not an Air Force vs. Navy issue. Both the Air Force and the Bureau of Aeronautics want nuclear-powered airplanes if at all possible, and so long as the Atomic Energy Act is in existence neither can hope to build a private empire in this field even if it so desired. The issue, and the controversy is really a fundamental one. It rests on the typical, perplexing, circular, hen and the egg, nature of all the decisions involving new weapons which the military are continuously asked to make. How valuable nuclear-powered aircraft might be depends heavily upon when it will be

(more)

available, what the maintenance problems are, what the cost will be in diversion of fissionable material, not only for the aircraft reactors installed, but also for the inventory required by the complex reprocessing procedures required for nuclear fuels. Unfortunately, the answers to these questions depend on the priority which is attached to the development program. How soon nuclear-powered aircraft can be available depends on how much effort we put into the program. Similarly, how soon we can give information as to probable performance and costs will depend on how rapidly the work progresses. The dilemma is very real and very serious. It was to aid in solving just such problems that the Weapons Systems Evaluation Group, under General Hull, was set up in the National Military Establishment.

The best summary of the situation which I can give is that the pessimists, who in general are those best informed, have thought through the immediate reactor and power-transfer problems and are staggered by the maintenance and operations problems which would be involved if the actual aircraft propulsion devices for combat use were to be based on our present knowledge and practices. The optimists on the other hand either are not yet aware of the very real immediate difficulties, or they are betting heavily on new ideas and new developments arising during the course of the work which avoid some of the currently foreseeable troubles. In this connection, I seem to recall, however, that not many years ago all the technical facts, and all the arguments of the experts, indicated an upper limit of 100,000 pounds as an absolute ceiling for the size of heavier-than-air-craft. This is an area in which it will no doubt be wise for us to be open-minded but skeptical. The best we can hope for in a program such as this is one in which the best available advice is sought and used.

In my introduction I raised the question as to whether our objective was worth while and whether we had a sound approach. Let me now try to summarize the situation as it looks at the present time.

We want to maintain technical leadership in the atomic energy field. This is our objective and our assignment by Act of Congress. As part of this effort we want a vigorous reactor program. This program must earn its keep for either peace or war purposes. The initial part of this program is the same whether the ultimate use of the reactors developed is civilian or military. We have a generously supported applied research program in the large national laboratories, to give us basic information for new developments and for providing a trained cadre of specialists in the atomic energy field. We have engineering task forces attacking some of the most promising

(more)

possibilities available to us at the present time. Most urgent are the two premium fuel uses of interest to the military, namely, power for ships and power for aircraft. The first of these while difficult, can conceivably be attained by direct frontal attack. The second is being attacked indirectly with the Materials Testing Reactor, representing an important, exploratory advance as well as providing an almost essential research tool. The MTR will be to reactor development what wind tunnels are to aircraft developments.

Another strong task force is engaged in a difficult but promising assignment on a reactor which can either be considered as giving power with fissionable material as a by-product or fissionable material with power as a by-product. In either case success would represent a major step in advance toward economical power for either military or civilian use.

The fourth task force is engaged in a frontal attack on the problems presented by the chronically short supply of fissionable material. Ideal success would increase by a factor of 139 the potential stockpile of fissionable material and might bring atomic energy for civilian use within sight. Even very partial success might go far toward helping us increase the efficiency of present production processes.

Success in all of these task force efforts is probably too much to hope for, but the possible return in each appears high enough so that success in one will pay for the rest. The risks are great but the stakes seem greater.

Details of the program are controversial and on these it is undoubtedly discreet for me to maintain a studied silence. I might be permitted, however, to end with a quotation from one of the wisest of scientists, Benjamin Franklin, which dates from the year 1780:

> "The rapid progress true science now makes, occasions my regretting sometimes that I was born so soon. It is impossible to imagine the height to which may be carried, in a thousand years, the power of man over matter... Oh that moral science were in as fair a way of improvement."

- 30 -

APPENDIX

COMPREHENSIVE TECHNICAL REPORT GENERAL ELECTRIC DIRECT-AIR-CYCLE AIRCRAFT NUCLEAR PROPULSION PROGRAM BY GUNNAR THORNTON JUNE 28, 1962

UNCLASSIFIED

1. INTRODUCTION AND PROGRAM BACKGROUND

1.1 INTRODUCTION

The General Electric Aircraft Nuclear Propulsion Program started in 1951 following feasibility studies performed primarily by the NEPA Project of the Fairchild Engine and Airplane Company. The work was performed under simultaneous contracts with the United States Air Force and the United States Atomic Energy Commission.

Air Force Contract

Air Force Contract No. AF33(038)-21102 required that the General Electric Company should accomplish "a development program and the manufacture and ground testing of a nuclear power plant suitable for testing at the earliest feasible date." In this connection, it was expected that there would be "a series of nuclear power plants" each of which was to "employ turbojet engines operated in conjunction with the selected nuclear reactor." It was also stated that "successive power plants in the series shall have improving performance to the end that the final one fabricated under this contract shall be suitable for testing in a military prototype nuclear aircraft."

In addition, the General Electric Company was to "carry out secondary work including theoretical, analytical, design and experimental studies on alternate types of propulsion systems and components in order to determine their relative merits for future development."

AEC Contract

The Atomic Energy Contract No. AT(11-1)-171 stated that "It is the objective of the Commission to develop, within the shortest practicable time, a nuclear reactor which, in conjunction with propulsion equipment, will fulfill the Air Force's requirements for the propulsion of aircraft." It was expected that "Attainment of this objective will require the erection and operation of several preliminary nuclear reactors."

More explicitly, General Electric would undertake "such research and development work on reactors as may be necessary or desirable to establish suitable design and specifications" and would also undertake "fabrication, assembly and testing of reactors and their component parts."

Furthermore, "secondary work" was to be performed on "alternate types of nuclear reactors" consisting of "theoretical, analytical, design, and experimental studies to determine the merits of the alternate types for future development."

The program proceeded in accordance with these contractual statements until its termination in March of 1961.

Scope of the General Electric Program

As required by the AEC contract, a series of preliminary reactors were developed and tested. The first operation of an aircraft engine on nuclear power was achieved on January 31, 1956, using an experimental direct-air-cycle reactor and a modified General Electric J47 turbojet engine. This was followed by a series of additional reactor operations using im-

proved reactor designs and materials. Concurrently, high performance turbomachinery (X211) was under development which could be used for a variety of nuclear propulsion system applications at both subsonic and supersonic speeds.

A series of power plants was designed combining the turbomachinery and the continually improved reactor materials and components. Several power plants suitable for test and operation in military aircraft were designed to meet specific military objectives. Each of the objectives was withdrawn, in accordance with the constantly evolving requirements of the national defense establishment, prior to final fabrication and operation of a prototype unit.

The power plant under development at program termination, the XNJ140E nuclear turbojet (Figure 1.1) was designed in accordance with Department of Defense guidance for a nuclear propulsion system capable of propelling a Convair NX2 (Figure 1.2) or equivalent aircraft at high subsonic speeds for 1000 hours before refueling. This was equivalent to a total range of approximately 500,000 miles, exceeding the total range of equivalent fully loaded chemical aircraft by a factor of approximately 100. An aircraft with this capability was believed to be best suited for an airborne alert and counterstrike mission in which it would remain airborne for periods of five days at a time, carrying ballistic missiles with nuclear warheads for air launch from outside the target area.

Growth versions of the XNJ140 power plant were in preliminary design for use in reconnaissance-counterstrike or similar missions at speeds of Mach 2.5 or greater. Advanced design and development of nuclear turbojets, ramjets, and rockets to meet other potential military objectives had been completed or were in process for both subsonic and supersonic applications.

Program Termination

The Aircraft Nuclear Propulsion program was terminated following the President's annual budget message to Congress on March 28, 1961, recommending omission of funds for program continuation. The program termination was based primarily on the fact that there was not considered to be a specific military requirement for a manned aircraft with the characteristics of the subsonic, long endurance system that was under development. The work on alternative subsonic missions and on growth versions for supersonic operation was simultaneously discontinued. The work on the unmanned nuclear ramjet and nuclear rocket propulsion continued in the national laboratories.

This summary volume of the Comprehensive Technical Report describes the program background, gives a brief physical description of the major power plant designs and test assemblies, and traces the development progression in basic technology to its final status. It also serves as a guide to the more detailed technical volumes of the Report (APEX-902 to 921 on the inside front cover), and to documents containing original source material.

Program background prior to initiation of the General Electric program is provided in the final report of the NEPA project (reference 1).

Matters of national policy concerning the program are discussed in transcripts of the 1959 Hearing of the Joint Congressional Committee on Atomic Energy and the report of the committee (references 2 and 3). The Congressional documents also enumerate various program evaluations which were performed by government committees.

The related activities of other contractors and government agencies participating in the aircraft nuclear propulsion program are described in reference 4 in the Congressional

committee reports, and in the technical reports of the participating contractors. The presidential budget message which prompted termination of the program is given in the Congressional Record (reference 5). Further information concerning the program termination and disposition of its activities are given in the annual report of the Atomic Energy Commission for the year 1961 (reference 6).

The detailed progression of the General Electric program is provided in the Quarterly Progress Reports (reference 7) and in the Annual Program Reports (reference 8).

1.2 PROGRAM BACKGROUND PRIOR TO 1951

The Aircraft Nuclear Propulsion Program resulted from actions during and shortly after World War II by individuals who were interested in applying the developments of the Manhattan District Program to aircraft propulsion. In 1946, the Air Force established the Fairchild NEPA (Nuclear Energy for the Propulsion of Aircraft) Project to study the feasibility of nuclear powered flight. In 1948, the Atomic Energy Commission contracted with the Massachusetts Institute of Technology for a group of prominent scientists, assembled as the "Lexington Project," to review the work being conducted in the field of aircraft nuclear propulsion. The Lexington Project recommended an expanded evaluation program and predicted that nuclear flight could be achieved in a subsonic system in "approximately 15 years at a cost well in excess of one billion dollars" (reference 9). Subsequently, in 1949, the Atomic Energy Commission entered the program on an active basis in a joint effort with the Air Force, the Navy, and the National Advisory Committee for Aeronautics. The feasibility studies continued at NEPA and at the Oak Ridge National Laboratories. The NEPA Project continued under joint auspices until May 1951 when the study was closed out on the basis that general feasibility had been established. The NEPA Final Status Report (reference 1) describes the early program background and the results of the studies that were performed.

1.2.1 MOTIVATION FOR NUCLEAR FLIGHT

The following statements, extracted from the NEPA report, define the motivation for establishing a program for the development of nuclear flight.

"The present air strategy of this country is defined by the speed, altitude, and range at which its aircraft can fly. As long as it is necessary to use conventional chemical fuels such as gasoline or kerosene, aircraft will be subject to limitations as to their characteristics of range, speed, and altitude. With chemical fuels, the factors of speed and range are incompatible...

"It is in the combination of high speed with long range that nuclear power promises to achieve results unobtainable by any other means... the only practical limitations on the range of a nuclear-powered aircraft would be the endurance of the crew against both radiation and ordinary fatigue, and the freedom of the power plant and aircraft from mechanical breakdown or battle damage.

"... For example, a nuclear-powered aircraft could:

- Not only fly to any point on the globe, but do so at very high speed, and by any route whatever, no matter how circuitous, and could return at high speed by an entirely different route. This is impossible for conventional aircraft.
- Fly at very low altitudes where radar detection is difficult and at high speeds for essentially unlimited distances, a performance impossible for chemically fueled aircraft.

- Fly completely around the earth at a speed equal to the easterly speed on the earth's surface at latitudes of military interest, thus conducting its mission entirely in darkness and hence with much reduced danger of enemy interference.
- Perform long-range reconnaissance missions or patrol for enemy submarines or aircraft for several days without stopping.
- Serve as an airborne communications center, particularly useful in polar areas, or as a mobile extension of the radar screen, or as a mobile airborne command post for large-scale military operations."

1.2.2 CONCLUSIONS OF NEPA PROJECT

It was the conclusion of the NEPA Project that nuclear powered flight was feasible. The following more specific conclusions (reference 1) of the NEPA Project provided the basis for the original orientation of the GE-ANP program.

- "A nuclear power plant can be developed which will be capable of propelling an inhabited bombardment aircraft at Mach 1.5 at 45,000 feet. The aircraft would be about the size of the largest present-day bombers. Its range to all intents and purposes would be unlimited."
- "Nuclear power plants to propel aircraft at subsonic speed and altitudes up to 35,000 feet can be built significantly sooner than those for Mach 1.5 and 45,000 feet... It may therefore be desirable to select as the first tactically useful power plant for development one designed for subsonic speed and altitude 35,000 feet or less..."
- "The most suitable type of propulsion machinery is the turbojet. Ducted fans and propellers are less desirable, but possibly of limited value... Nuclear ramjets and nuclear rockets may possibly become of interest in the distant future."
- "The air-cooled hydrogenous-moderated reactor is particularly attractive because of its relative simplicity and low vulnerability, but suffers from major uncertainties regarding the retention of fission products within the fuel elements and leakage of neutrons through the air ducts in the shield... The major uncertainties of the liquid-metal cycles are those of reliability and the degree to which the effects of a system leak can be controlled... Choice of any cycle for exclusive or even major development at this time would be arbitrary."
- "Further progress in the development of a nuclear power plant for an aircraft will require that the program contain a rapidly increasing proportion of the designing, building, and testing of full-scale power-plant components and assemblies. Basic investigation of the problems encountered, and development of fundamental supporting data, must also continue. New explorations, particularly in materials and metallurgy required for Phase II* performance, should be carried forward.
- "An accurate evaluation of the true worth of nuclear-powered flight will not be possible until a considerable amount of ground operation and flight experience is gained with this new propulsive system. In spite of the difficult nature of the development program and in spite of the inherent hazards of radiation, the potentialities of nuclear-powered flight are so great that its continued development, at least to the point where adequate ground and flight experience can be gained in one or more experimental articles, is mandatory in the interest of national defense."

*Defined in the NEPA report as "Tactical aircraft with the following design points ... : 60,000 feet at Mach 0.9; sea level at Mach 0.9; 35,000 to 45,000 feet at Mach 1.5; and 35,000 feet at Mach 2.0 ..."

1.3 HISTORICAL SUMMARY OF THE GENERAL ELECTRIC PROGRAM

The Air Force and Atomic Energy Commission agreed, in early 1951, that feasibility had been sufficiently well established to warrant initiation of active research and development leading to militarily useful aircraft nuclear propulsion systems. General Electric was selected as the propulsion system contractor on the basis of its experience in the development of both nuclear reactors and aircraft turbomachinery. The Air Force contract with the Aircraft Gas Turbine Division of the General Electric Company, to undertake active development of a nuclear turbojet engine, became effective on March 21, 1951. The Atomic Energy Commission contract for reactor development became effective on June 29, 1951.

The General Electric Program encompassed:

1. An applied research program in materials, engineering physics, and component development to provide a basic technology applicable to a broad spectrum of potentially useful aircraft nuclear propulsion systems
2. The design and test of experimental nuclear reactors operating aircraft turbomachinery
3. The design and test of advanced turbomachinery
4. The design and development of prototype propulsion systems to meet specific military objectives
5. The advanced design and development of propulsion systems to meet anticipated military requirements

1.3.1 EARLY FLIGHT PROGRAM

The first several months of the General Electric program were devoted to the selection of a system for initial development. This phase of the program was completed on August 28, 1951 with a recommendation to the Air Force and Atomic Energy Commission to proceed with the development of a direct-air-cycle system using an air-cooled reactor with metallic fuel elements and a hydrogenous liquid as the moderator and structural coolant. This recommendation was made on the basis that it represented the best way to obtain early nuclear flight experience. It was anticipated that higher performance materials would be required for ultimate application to operational military aircraft.

An Air Force objective was established for nuclear ground and flight operation in a modified Convair aircraft at the earliest feasible date in order to evaluate the operational practicability of nuclear systems prior to commitment to a prototype military nuclear aircraft. Approval was granted to proceed with the development of a power plant, designated the P-1, to meet the early flight objective. The initial ground test was scheduled for 1954 and flight test for 1957. The early flight objective was withdrawn in March 1953 on the basis that early flight demonstration with a system not fitting a specific military requirement was no longer considered warranted.

1.3.2 APPLIED RESEARCH AND DEVELOPMENT

After discontinuation of work on the P-1 power plant, the program was redirected to an applied research and development program applicable to a broad spectrum of potentially useful military propulsion systems. The applied research activity continued until program termination, developing the basic materials and engineering analysis methods used in a series of subsequent reactor operations and power plant designs. The materials program encompassed the development of metallic and ceramic fuel elements, hydrided metallic moderators, and shield, controls, and structural materials for use in both subsonic and supersonic aircraft. This was supported by extensive in-pile test programs. Engineering analysis techniques were developed in reactor and shield nuclear physics, aerothermodynamics, controls, mechanics, and nuclear safety. These activities were supported by a

strong experimental program including the performance of reactor critical experiments, shielding experiments, aerothermodynamic tests, and mechanical testing in simulated environments and in nuclear test reactors. The applied research activity is discussed in section 7 of this volume.

1.3.3 HEAT TRANSFER REACTOR EXPERIMENTS

A series of experimental reactors were built and operated using materials and methods developed in the applied research activity.

Heat Transfer Reactor Experiment No. 1

The first reactor operated in the General Electric program was in Heat Transfer Reactor Experiment No. 1. This was a direct-air-cycle reactor using nickel-chromium, uranium-oxide-dispersion fuel elements, with water serving the combined function of moderator and structural coolant. The HTRE-1 reactor first operated a modified General Electric J47 turbojet engine exclusively on nuclear power in January 1956. Operation of the HTRE-1 continued throughout the Calendar Year 1956, accumulating a total of 150.8 hours of operation at high nuclear power levels, exceeding the design requirement of 100 hours.

Heat Transfer Reactor Experiment No. 2

The HTRE-2 reactor was a modification of HTRE-1, providing a hexagonal center hole, 11 inches across flats with an active length of 30 inches, for use in testing insert sections for advanced reactors. HTRE-2 operation started in July 1957 and continued during the remainder of the program, accumulating 1299 hours of high power nuclear operation. Insert test sections consisted of metallic fuel elements combined with air-cooled hydrided zirconium moderators and beryllium oxide fuel elements for use in ceramic reactors. Inserts were operated at materials temperatures up to 2800°F for extended periods and for short periods at higher temperatures.

Heat Transfer Reactor Experiment No. 3

The HTRE-3 reactor was built in a full-scale aircraft reactor configuration using Ni-Cr fuel elements of the HTRE-1 type and an air-cooled hydrided zirconium moderator. Two modified J47 turbojets were operated by the reactor with full nuclear power being achieved in 1959. The system operated for a total of 126 hours; the design objective was 100 hours operation.

Proposed Heat Transfer Reactor Experiment No. 4

A ceramic reactor using beryllium oxide fuel tubes was designed and received extensive component development for a proposed fourth Heat Transfer Reactor Experiment in the HTRE-1 or HTRE-3 test assemblies. Consideration of HTRE-4 was dropped in favor of proceeding directly to a prototype propulsion system incorporating the ceramic reactor design features and components which had been developed.

1.3.4 TURBOMACHINERY DEVELOPMENT

A propulsion system is characterized primarily by the propulsion machinery; the heat source, chemical or nuclear, plays a secondary role. By 1955, sufficient progress had been made to define the characteristics of a basic turbojet propulsion unit into which nuclear reactor heat sources of successively higher performance capabilities could be incorporated with minimum modification to the turbomachinery. Development of this unit, the X211 turbojet engine, began in 1955. The X211 was a single-rotor, variable-stator, high-pressure-ratio engine with an airflow of approximately 400 pounds per second at sea level static; it had growth potential to turbine inlet temperatures above 2000°F for supersonic operation.

1.3.5 PROTOTYPE PROPULSION SYSTEMS

Prototype propulsion system designs all used X211 turbomachinery with reactors that had been tested or were planned for development in the reactor program. The power-plant configurations, performance requirements, and materials selection were based on specific military objectives or anticipated future requirements. Two prototype power plants, the XMA-1 and the XNJ140E, were designed to meet specific military objectives.

XMA-1 Power Plant

A specific military objective requiring a nuclear propulsion system was issued by the Air Force on March 22, 1955, as Systems Operational Requirement No. 81. This required a "piloted nuclear powered intercontinental strategic bombardment weapon system" capable of extended cruise without in-flight refueling, penetrating enemy defenses at high altitudes and supersonic speeds, and low-level attack at subsonic speeds. The design and development of a power plant designated the XMA-1 was undertaken to meet these requirements, assuming a nuclear cruise and chemically augmented sprint. The XMA-1 combined two sets of X211 turbomachinery with a single reactor. Initial ground test was scheduled for 1959 and the first flight test for 1960. A decision was made, late in 1956, to de-emphasize aircraft development but to continue developing the propulsion system at a reduced level without reference to a specific military objective.

A new objective was provided by the Air Force on October 28, 1958, as Systems Operation Requirement No. 172 for "a Continuous airborne missile launcher and low level weapons system" (CAMAL). The CAMAL mission retained the extended cruise and low-level penetration of SOR 81 but substituted the use of long-range, air-to-ground ballistic missiles for the high altitude supersonic portion of the flight regime. The XMA-1 development was redirected toward the CAMAL objective. First flight in 1963 was assumed as a target date using an early model of the power plant, the XMA-1A with a reactor of the type tested in HTRE-3. An improved model, the XMA-1C, using an advanced metallic or ceramic reactor was placed in preliminary design. The proposed Convair Model 54 aircraft was considered to be the flight vehicle.

XNJ140E Power Plant

In 1959, the early flight objective for the CAMAL aircraft was eliminated. In place of a specific weapons system objective, general guidance was provided to direct the applied research and development program toward a propulsion system capable of propelling an aircraft at a speed of Mach 0.8 to 0.9 at an altitude of approximately 35,000 feet. The reactor was to be capable of 1000 operating hours at the specified performance level and was to have development potential for even higher performance. In view of the revised objective, work on the XMA-1A reactor was discontinued.

After an evaluation of the relative development status of high temperature metallic and ceramic materials that had been under development for the XMA-1C reactor, a BeO ceramic reactor was selected to meet the Department of Defense guidance. Simultaneously, it was determined that a single-rather than dual-engine configuration was better suited to meet the growth potential requirement.

The development of such a power plant, designated tha XNJ140E, was proposed in March 1960 and was subsequently approved as a development objective. The XNJ140E used the basic components of the X211 turbomachinery and the reactor and shield materials that had been under development for the XMA-1C. A target date for ground test of a prototype unit, the XNJ140E-1, was set for December 1962. The ground test was referred to as the "Advanced Core Test," turbomachinery having been previously tested under chemical power.

On November 9, 1960, the Air Force issued Advanced Development Objective No. 20 defining the objectives of the Air Force Nuclear Aircraft Development program. The immediate objective was to achieve nuclear flight in a military prototype aircraft at subsonic performance levels matching the Department of Defense guidance. Supersonic nuclear flight was the ultimate objective.

The initial system would be used to evaluate the practicality of subsonic nuclear missions of long endurance, such as air alert, missile launching, low-level penetration, logistics, reconnaissance, air early warning, antisubmarine warfare, and airborne command posts. The subsonic system would also serve to develop a basic equipment and operational technology leading toward supersonic nuclear aircraft capability.

In accordance with the ADO No. 20 objective a target date was established to furnish XNJ140E power plants for initial flight operation in the Convair NX2 aircraft in 1965 after completion of ground testing of the XNJ140E-1.

1.3.6 ADVANCED PROPULSION SYSTEMS

In addition to the design of power plants to meet specific military objectives, propulsion systems were being designed to meet anticipated future military requirements. Primary emphasis was placed on a power plant capable of propelling a B-70 type aircraft at a speed of Mach 2.5 and an altitude of 45,000 feet. The basic XNJ140 power plant configuration was used in these studies, with advanced versions of the X211 turbomachinery designed to higher temperature capability. Smaller reactors of higher temperature capability were under development for the supersonic application. Additional design studies were being made on power plants of even higher performance capability and on propulsion systems for subsonic, load-carrying aircraft, closed-gas-cycle systems, nuclear ramjets, and nuclear rockets.

1.3.7 STATUS AT PROGRAM TERMINATION

The XNJ140E program was on schedule when the Aircraft Nuclear Propulsion Program was terminated. The basic propulsion machinery had been operated for a total of 758 hours with a chemical heat source, using several engine buildups. Structural modifications were in process to adapt the turbomachinery to the final configuration of the XNJ140E reactor-shield assembly. Design of the reactor-shield assembly and subassembly had been completed. Manufacturing drawings of individual parts and components were completed or in process. Prototypes of critical components had been proof-tested. Beryllium oxide fuel tube assemblies for use in the reactor had been tested for a total of 10,683 hours in the MTR, ORR, and as inserts in an experimental aircraft reactor at temperatures approximating or exceeding design requirements. Reactor critical experiments had been performed to verify fuel element loading specifications. Permission had been requested to proceed with manufacture and assembly of the reactor into the propulsion machinery. Ground test operation was scheduled for late 1962. Nuclear flight test was predicted in a test bed aircraft for 1963, and nuclear flight operation of the NX2 prototype military aircraft in 1965.

APPENDIX

Total ANP Program Costs
(in $ millions)

AGENCY	1946-1951	1952-1961	TOTAL
Atomic Energy Commission			
Direct Cycle Propulsion	0	299.2	299.2
Indirect Cycle Propulsion	7	173.0	180.0
Prelim. Studies & Support	0.4	28.0	28.4
Total			507.6
USAF			
Direct Cycle Propulsion	0.4	261.6	262.0
Indirect Cycle Propulsion	0	98.5	98.5
Airframe & components	0.2	117.2	117.4
Feasibility Studies & Support	19.5	21.3	40.8
Total			518.7
US Navy			
Direct Cycle Propulsion	0	1.5	1.5
Indirect cycle Propulsion	0	1.9	1.9
Airframe & Components	0	5.7	5.7
Feasibility Studies & Support	1.5	3.4	4.9
Total			14.0
Total Cost of ANP Program	28.9	1011.4	1,040.4

APPENDIX

Total ANP Program Costs by Prime Contractor
(in $ millions)

CONTRACTOR	OPERATING	FACILITIES	TOTAL COSTS
General Electric Co.	462.9	64.5	527.4
All other direct cycle Firms	5.1	30.3	5.4
Pratt & Whitney Div.	138.2	26.3	164.5
access roads constructed		.5	.5
Corps of Engineers (CANEL Facilities)		41.5	41.5
Union Carbide (at Oak Ridge)	67.7	0.6	68.3
Convair Div.	63.9	6.4	70.3
Lockheed Aircraft Co.	19.2	14.5	33.7
All other airframe contractors	19.1	0	19.1
(Feasibility Studies)			
Fairchild Corp.	21.0		21.0
Union Carbide (at Oak Ridge)	23.5	1.4	24.9
Other contractors	19.0	14.7	33.7
TOTAL PROGRAM	839.7	200.7	1,040.4

APPENDIX

87TH CONGRESS, 1st Session } HOUSE OF REPRESENTATIVES { DOCUMENT No. 123

RECOMMENDATIONS RELATING TO OUR DEFENSE BUDGET

MESSAGE

FROM

THE PRESIDENT OF THE UNITED STATES

RELATIVE TO

RECOMMENDATIONS RELATING TO OUR DEFENSE BUDGET

MARCH 28, 1961.—Referred to the Committee on Appropriations and ordered to be printed

---

To the Congress of the United States:

In my role as Commander in Chief of the American Armed Forces, and with my concern over the security of this Nation now and in the future, no single question of policy has concerned me more since entering upon these responsibilities than the adequacy of our present and planned military forces to accomplish our major national security objectives.

In January, while ordering certain immediately needed changes, I instructed the Secretary of Defense to reappraise our entire defense strategy, capacity, commitments and needs in the light of present and future dangers. The Secretary of State and others have been consulted in this reappraisal, and I have myself carefully reviewed their reports and advice.

Such a review is obviously a tremendous task and it still continues. But circumstances do not permit a postponement of all further action during the many additional months that a full reappraisal will require. Consequently we are now able to present the most urgent and obvious recommendations for inclusion in the fiscal 1962 budget.

5. Nearly 15 years and about $1 billion have been devoted to the attempted development of a nuclear-powered aircraft; but the possibility of achieving a militarily useful aircraft in the foreseeable future is still very remote. The January budget already recommended a severe curtailment of this project, cutting the level of effort in half by limiting the scope to only one of the two different engines under development, although not indicating which one. We believe the time has come to reach a clean-cut decision in this matter. Transferring the entire subject matter to the Atomic Energy Commission budget where it belongs, as a nondefense research item, we propose to terminate development effort on both approaches on the nuclear powerplant, comprising reactor and engine, and on the airframe; but to carry forward scientific research and development in the fields of high temperature materials and high performance reactors, which is related to AEC's broad objectives in atomic reactor, development including some work at the present plants, making use of their scientific teams. This will save an additional $35 million in the Defense budget for fiscal 1962 below the figure previously reduced in January,

CHRONOLOGICAL HISTORY

OF

NEPA

Nuclear Energy for Propulsion of Aircraft

and

GE-ANP

General Electric Aircraft Nuclear Propulsion Program

1946 - 1961

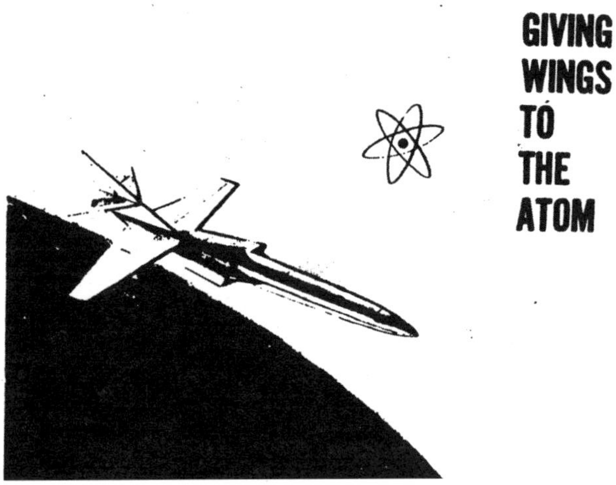

ATOMIC PRODUCTS DIVISION

GENERAL ELECTRIC

AIRCRAFT NUCLEAR PROPULSION DEPARTMENT

CINCINNATI 15, OHIO

This booklet was prepared as a memento of the 45th Anniversary Reunion of POPSEE August 21, 1996

CHRONOLOGICAL HISTORY OF NEPA AND GE-ANP

1941 - An early champion of nuclear flight was Col. Donald Keirn, an Air Force power plant specialist at Wright Field in Dayton, Ohio. Keirn went to England to consult with Sir Frank Whittle, Britain's jet engine pioneer, and returned with details of Whittle's engine. He then acted as liaison between the Air Force and General Electric in producing the first U.S. jet engine. In the course of developing the first jet engine, Keirn got to know D.R. Shoults, then an engineering executive at General Electric Co. who became responsible for coordinating the GE efforts with aircraft manufacturers. Keirn and Shoults spent time together discussing linking the jet engine with nuclear power for the propulsion of aircraft. Shoults left GE after the War for a different industrial employer, but in the next decade came back to GE to be Manager of the GE Aircraft Nuclear Propulsion Program. Keirn continued to pursue support from other sources.

July 1945 - Shortly before the first nuclear explosion at Alamogordo, New Mexico, Col. Keirn, Chief of Power Plant Lab at Wright-Patterson Air Force Base requested information as to possible application of atomic energy to aircraft from Dr. Vannevar Bush of the Manhattan District. Bush advised Keirn to defer such application because efforts were directed to urgent development of an explosive.

Aug. 1945 - Smyth Report gave public some knowledge of atomic bomb and the potential of atomic energy. Gordon Simmons Jr., a young engineer involved in construction of K-25 plant at Oak Ridge, addressed letter to Sherman Fairchild stating that he would like to be connected with a company interested in applying nuclear energy to propulsion of aircraft. His thoughts had been stimulated by J.Carlton Ward Jr., President of Fairchild Engine & Airplane Corp., who had expressed the strategic advantages to be gained by harnessing nuclear energy to power large aircraft.

Oct. 1945 - Conference held in Fairchild's New York office at which Fairchild decided to make presentations to the military services for sponsorship of such a project. After numerous conferences and consultations, the Army Air Force decided to sponsor a single unified project with NACA participating.

Jan. 1946 - Meeting held at Pentagon of all interested agencies and aircraft engine companies; after concurrence, the Air Force asked the industry members to select one of their companies as the single manager of the group effort and the prime contractor to the Air Force. Fairchild Engine & Airplane Corp. was selected as the leader.

May 28, 1946 - NEPA Project began officially with signing of AAF letter of intent by General Spaatz and Maj. Gen. Leslie Groves. Associated with Fairchild were Allison, United Aircraft, Wright Aeronautical, General Electric, Westinghouse, Continental Aviation, Lycoming, Northrop, Flader and Menasco Mfg. Co. Member companies were to have a voice in technical phases of the NEPA feasibility investigation and research leading to the adaptation of nuclear energy as a means of propulsion of aircraft for tactical utility. Letter-contract authorized expenditure of $200,000; subsequent 18 amendments increased funding to $5.25 million.

July 15, 1946 - NEPA Project offices moved into suite of rooms in Raleigh Hotel on West 72nd St. New York City.

Aug. 30, 1946 - NEPA Report No. 2 recommended construction and testing a complete nuclear power plant and the continuing development of a high temperature reactor.

Sept. 4, 1946 - NEPA personnel moved to Oak Ridge, Tenn. occupying space in AEC Administrative Building.

Sept. 24, 1946 - NEPA offices moved to S-50 Area near K-25 Power Plant. (Previously used for pilot operation of a uranium-enrichment technology known as S-50 Thermal Diffusion, which used high pressure steam to separate the isotopes of uranium for use in atomic bombs). This move gained proximity to ORNL experimental facilities and access to Manhattan District classified reports. NEPA secured the part time services of a team of consultants including eminent nuclear specialists.

Nov. 1946 - Hiring began, but moved slowly because Oak Ridge housing quota for NEPA was limited. (Oak Ridge was still a closed city.) The constant crises concerning the continuation and future of NEPA and resulting hiring moratorium created job insecurity in minds of many technical candidates.

Feb. 1947 - NEPA came under AEC Subcommittee of the Joint Research & Development Branch.

Mar. 1947 - Important milestone passed: NEPA staff recommended major effort concentrate on direct air-cooled ceramic reactor used with a turbo-jet engine. Member companies and government agencies concurred.

Oct. 1947 - NEPA Staff undertook a design study of a supersonic turbojet missile; this study continued to Summer 1948.

Dec. 1947 - R & D Board of Defense Dept. recommended that NEPA Program proceed on priority basis as a coordinated project with AEC. NEPA personnel totaled 264 of which 107 were technically trained. Due to lack of laboratory facilities, materials work was confined to literature surveys, particularly high temperature alloys and ceramics.

Jan. 1948 - Finletter Report recommended intensifying research efforts on nuclear plane.

Mar. 1948 - Congressional Aviation Board urged NEPA be given highest priority in atomic energy research.

June 1948 - Lexington Project created by contract between AEC and MIT with Walter Whitman as director.

Sept. 1948 - Lexington Report predicted a nuclear powered aircraft feasible and could be achieved in 15-20 years at cost of $2 billion.

Nov. 1948 - Materials Laboratory completed in remodeled Building F-10.

Dec. 1948 - Total NEPA personnel 444 (164 technical).

Feb. 1949 - Nuclear Powered Flight Program became joint effort of Air Force, Navy, AEC and NACA, implemented by establishment of an Aircraft Nuclear Propulsion group at ORNL with prime responsibility for reactor and shield.

May 1949 - Medical unit established as NEPA Medical Department to provide routine dispensary care and physical examinations. Dr. Buyniski began investigations of diagnosis of beryllium poisoning and exposure to combustion products of liquid metals, lithium and sodium.

July 1949 - Industrial Hygiene Control Unit established to monitor fabrication of beryllium and beryllium carbide as well as radiation safety in laboratories.

Oct. 1949 - NEPA personnel total 658 (270 were technically trained). Average salary of BS Degree with 10 years experience was $475 per month (slightly above national average). Starting salary for new graduates averaged $300. The cost of renting a three bedroom house in Oak Ridge was increased to $90 per month.

Oct. 1949 - An article in the Oak Ridger newspaper, reprinted from the Louisville Courier-Journal, described NEPA as one of the most interesting and least publicized Oak Ridge activity. "Most of the 600 employees are persons of high technical or scientific rank, seeking to find some way to use atomic energy to make airplanes fly".

During 1949 the greatest achievement was the divided shield concept (shielding of the crew compartment) which reduced gross weight by 30% to 650,000 pounds. Studies were also conducted on binary bismuth cycle, ternary liquid metals cycle, helium compressor jet and turbo-jet cycles, fabrication of beryllium carbide bodies, solubility of container materials in various liquid metals, and testing of liquid metals handling in circulation rigs. Plans were made for Tug-tow Tests by the Air Force. Design of a critical experiment facility was completed and some components built and tested. A small scale air cycle powerplant was constructed using a turbojet engine and electrical heat source. Circulating fuel type reactors were studied and posed new materials and shielding problems. A survey of jet engine manufacturers was made to establish limitations on engine size and characteristics.

Nov. 1949 - AEC began ANP research project at ORNL.

Jan. 1950 - Concept of water-moderated reactor patented by Thornton and Corbin made operation of a reactor in an aircraft feasible.

Mar. 1950 - Aircraft Reactors Branch established in AEC; assumed responsibility for AEC part of NEPA.

Aug. 1950 - NEPA personnel peaked at 676 (18 PhD, 49 MS, 128 BS). General Manager was T.A. Sims Jr.; Director of Technical Division was M.C. Leverett. Technical Director and Chief Engineer was A. Kalitinsky. Several NEPA people were assigned to work with ORNL on the design of the Aircraft Reactor Experiment (ARE). The Critical Experiment Building was completed and NEPA test equipment was installed.

Nov. 15, 1950 - Fairchild entered into contract with AEC for work related to nuclear aspects of the then current Aircraft Nuclear Propulsion Program (ANPP). This resulted in modifications to Fairchild-Air Force prime contract. (Almost simultaneously Fairchild was advised of Air Force - AEC decision to consider first phase of ANP program at an end).

Jan. 1951 - Air Force decided that the program had advanced to the stage of "hardware development" and the stage of feasibility studies should be closed. The mission of the NEPA project had been accomplished. The hardware development phase would be carried out by the General Electric Company and responsibility for the aircraft was placed with Consolidated Vultee Aircraft Company.

April 30, 1951 - Termination date for NEPA Project. At this date a formal cycle selection had not been made. The two major cycles of interest were the liquid-coolant (sodium) cycle and the direct air cycle. During the five year life of the Fairchild NEPA contract, total funding reached $23.84 million; (40.3% for salaries and wages, 24.2% for subcontractors). The high average quality in both experience and performance of NEPA personnel was attested by the fact that 38 separate industrial companies and research laboratories made vigorous efforts to employ NEPA personnel at the time of termination. Many technical people had numerous offers from which to choose.

May 1, 1951 - General Electric initiates ANP program in Aircraft Gas Turbine Division (AGT) in Cincinnati aimed at development of power plant for eventual flight test. Eighty-seven NEPA project people joined GE and continued work at Oak Ridge while facilities at Evendale, Ohio were prepared for the nuclear research. D.R. Shoults was Project Manager and M.C. Leverett was Manager of Engineering. First six months' effort devoted to evaluation of the two cycles. In October the direct air cycle was selected for development and work on liquid metals was phased out at GE. Total personnel 450.

July 1951 - Fuel sheet composed of uranium dioxide-stainless steel core with stainless steel cladding was successfully fabricated. This metal clad cermet core

fuel sheet was the basis for the direct cycle reactors used in all High Temperature Reactor Experiments (HTRE-1, HTRE-2 and HTRE-3) tested in Idaho.

May 1952 - AEC approved use of part of National Reactor Testing Station (NRTS) at Arco, Idaho as flight test base.

July 1952 - AEC and Defense Dept. informed Joint Committee on Atomic Energy (JCAE) that plans were being made for flight test of a nuclear propulsion system in the 1956-58 period utilizing a modified B-36 as the test bed.

Aug. 1952 - Oak Ridge operations and personnel moved to Cincinnati (Bldg. D).

Dec. 1952 - Office for ANP established to coordinate AEC and Air Force participation in program. Maj. Gen. Donald Keirn named director.

Jan. 1953 - Eisenhower Administration began. Charles Wilson named Secretary of Defense.

Mar. 1953 - Air Force Scientific Advisory Board recommended cutting back ANP by 50 percent on grounds that activities unwarranted by state-of-the-art and rate of progress.

April 1953 - National Security Council ordered AEC and Defense Dept. to cancel ANP Program on grounds of budget savings and program not in national interest. Secretary of Defense Wilson termed the nuclear plane a "shitepoke" and ordered the program canceled. Wilson defined a "shitepoke" as "a great big bird that flies over the marshes, does not have much of a body or speed to it or anything, but it can fly".

May 1953 - Joint Committee called for meeting with Secretary of Air Force Talbott and Deputy Secretary of Defense Keyes. "Cancellation" of project termed misinterpretation of order. Reorganization of project underway. ANP Program redirected toward applied research and development on limited funds basis. A series of high temperature reactor experiments (HTRE) were scheduled to develop and prove-out the reactor power plant. (Personnel total 1300).

Sept. 1953 - Aircraft Nuclear Propulsion Department (ANPD) established under GE Atomic Products Division.

Dec. 1953 - Air Force informed AEC of its renewed interest in manned nuclear aircraft and asked AEC to expedite experimental work.

Jan. 1954 - GE Savings & Stock Bonus Plan initiated with U.S. Savings Bonds having six year maturity.

April 1954 - Director of ANP Project, Maj. Gen. Keirn, advised Joint Committee that nuclear aircraft could be in operation in half scheduled time if given high priority. Joint Committee approved report by R & D Subcommittee calling for "crash" effort on ANP Project. Report sent to President Eisenhower, Secretary of Defense and Chairman of AEC.

June 1954 - GE-ANP personnel reduced by 40 percent to about 800.

July 1954 - Joint Pratt & Whitney and ORNL program established to develop indirect liquid metal cycle propulsion system.

July 1954 - The concentric ring fuel element design was proven successful using fuel sheet of Nichrome clad uranium dioxide cermet core. This design was the basis for all HTRE reactors.

July 1954 - Fabrication techniques were developed establishing hydrided zirconium as a practical solid moderator material. This simplified reactor design, and the material was used in the High Temperature Reactor Experiment No. 3 (HTRE-3).

Aug. 1954 - Critical experiments were initiated at Evendale to provide data for the design of HTRE-1.

Oct. 1954 - Idaho Test Station (ITS) achieved Section status in ANPD. Manager of the ITS ("Remote Site") was W.H. (Bill) Long.

Oct. 1954 - Air service for GE-ANP personnel to ITS was initiated by the Air Force. Later, a C-46 was retrofitted with passenger seats and a galley for this "Site Flite" service. The non-stop flight from Wilmington OH to Idaho Falls was made in 8 hours.

Feb. 1955 - AEC reported progress on direct cycle reactor exceeded expectations and authorized additional funds to be spent in fiscal 1955.

April 1955 - Air Force issued requirements for 125-A Weapons System high-performance, nuclear-powered aircraft and initiated program with project office at Wright Field, Dayton. Competition for airframe studies began.

June 1955 - AEC and Defense Dept. agree to accelerate ANP program with objective of testing prototype about 1959.

Sept. 1955 - Test aircraft flown with 3 megawatt reactor aboard a modified B-36 (called X-6), to measure radiation from reactor in flight. A total of 44 flights were made in two years from Carswell Air Force Base, Fort Worth, TX to Roswell, NM. where the reactor was powered-up over the desert area. A B-50 aircraft followed with the radiation instrumentation. (The B-36 is currently being restored at Fort Worth).

Sept. 1955 - Pratt & Whitney authorized to work on indirect (liquid metals) cycle reactor. Construction of CANEL Facility (Connecticut Aircraft Nuclear Engine

Laboratory) started to accommodate expansion of ANP Program. (Effort on indirect cycle was reduced to secondary status in Sept. 1957).

Nov. 1955 - Air Force directed team-up of General Electric with Convair and Pratt & Whitney with Lockheed to proceed with propulsion systems for high-performance aircraft.

Dec. 1955 - Yttrium oxide addition to uranium dioxide was found to stabilize the fuel in air to 2600 °F.

Jan. 1956 - GE-ANP personnel 1700. HTRE No.1 test operated on schedule in Initial Engine Test (IET) Facility in Idaho. This direct cycle water-moderated reactor with Nichrome clad fuel elements was used to power a turbojet engine successfully.

Mar. 1956 - Ceramic fuel elements composed of beryllium oxide-uranium oxide-yttrium oxide were successfully tested in-reactor at 2500 °F.

June 1956 - Air Force Chief of Staff General LeMay told Joint Committee he was interested in achieving nuclear flight at earliest practical date. Maj. Gen. Keirn said ground test possible in 1959 and first flight in 1960.

Aug. 1956 - Revised fiscal 1957 program resulted in 18 months slippage in program schedule. Defense Dept. policy decision cut back ANP Program.

Dec. 1956 - HTRE-2 test reactor operated successfully at Idaho Test Station. In the following two months of testing over 150 hours of nuclear powered turbojet engine operation were completed.

Dec. 1956 - Meeting of Defense Dept. and Budget Bureau officials with the President in Augusta GA eliminated effort on indirect cycle and reduced effort on direct cycle development.

Jan. 1957 - AF Scientific Advisory Board recommended less emphasis on engine and airframe development, more on reactor research and development.

Feb. 1957 - GE-ANP Program personnel total 2900.

Feb. 1957 - A test of 4000 ceramic tubular fuel elements in a special insert in HTRE-2 showed that water vapor corrosion of beryllium oxide was a major problem. (Note: This problem was later solved by a process of co-extruding beryllium oxide fuel elements with a coating of yttrium oxide stabilized zirconium oxide).

Feb. 1957 - Littlewood Committee of Defense Dept. began review of ANPP. Joint Committee called Defense Dept. and AEC officials to testify on status of ANP Program; urged efforts to achieve early flight for prototype nuclear system. Defense Deputy Secretary Quarles said no flight date would be set until propulsion system was developed adequate for military plane. Joint Committee sent letter to Quarles expressing concern about lack of firm program objectives and lack of direction. Defense Dept. appointed ad hoc panel of General Officers (Mills Panel) to review ANP Program and missions contemplated.

Feb. 1957 - GE Management Association Ladies Night Dance was held at Castle Farm with Lionel Hampton Orchestra providing music and entertainment for 1300 couples.

April 1957 - Joint Committee urged Defense Dept. to proceed with vigorous ANP program; met again with Secretary Quarles to emphasize its concern. Quarles testified that program objectives had been established for ground and flight-test propulsion systems with first flight in 1960. Littlewood Committee report recommended ANP development program be carried through flight test stage.

May 1957 - Budget Bureau sent directive to executive agencies requiring fiscal 1959 budget be held at same or lower level than FY58. Canterbury Panel of Air R & D Command recommended development of low-level nuclear plane.

June 1957 - Unified project direction under Maj. Gen. Keirn was formed to coordinate AEC and Defense Dept. activities. Mills Report recommended early fabrication and flight testing of prototype propulsion system in 1960's. Test aircraft to be immediate objective. Deputy Secretary Quarles approved Air Force program.

Sept. 1957 - A major increase in performance of metallic fuel element reactors resulted from finding that unclad hydrided zirconium moderator components were capable of operation in air at 1200 °F.

Sept. 1957 - Budgetary ceilings cause slippage in time schedule for ground test of direct cycle system. P & W efforts on indirect cycle reduced to low level.

Oct. 1957 - Soviet Sputnik launched into orbit.

Oct. 1957 - Hunsaker Committee established to review hazards of nuclear plane and recommend national policy. Rep. Price, Chm. R & D Sub-committee, after trip to Russia, sent letter to the President urging early flight program in view of Russia's success with Sputnik; expressed concern over lack of well defined objectives and target dates.

Nov. 1957 - GE-ANPD Organization with 3140 employees consisted of ten Sections under D. R. Shoults as General Manager.

Dec. 1957 - Production of high-purity yttrium metal with low thermal neutron absorption cross-section was begun in large quantities.

Dec. 1957 - Hunsaker Committee report expressed concern over potential hazards of nuclear flight over land; recommended test flights be made from island or coastal base. Joint Chiefs said such action was premature.

Jan. 1958 - President Eisenhower requests his science advisor, Dr. Killian, to review ANP program; Dr. Bacher appointed chairman of study committee.

Feb. 1958 - Bacher Committee recommended greater emphasis on advanced materials capable of producing higher performance reactor. With recommendation of Dr. Killian and Dr. Bacher, the President approved. Early flight proposal of Air Force was postponed.

April 1958 - Unclad hydrided yttrium used as a moderator component was shown to be stable in air to 1600 °F, thus promising increased performance for metallic fuel element reactors.

May 1958 - GE-ANP Spring Swing Dance held at Castle Farm; tickets $1.00 per person included refreshments and door prizes.

June 1958 - Air Force dropped 125-A Weapons System and replaced with CAMAL Mission (Continous Airborne Missile Launching and Low-level Penetration System). CAMAL more realistically reflected the "state-of-the-art" of the A-Plane, specifying low altitude bombing attack at subsonic speeds.

June 1958 - General Electric Theater presented on CBS Television network on Sunday evenings with Ronald Reagan as host and frequent star.

Aug. 1958 - GE Employees Activities Association (GEEAA) Recreational Park officially opened to GE employees and their families. Participating in ceremonies were ANP'ers Ray Currens and Gert Shuette.

Oct. 1958 - The development of a high performance ceramic reactor based on beryllium oxide was accelerated to meet longer range goals of the Air Force.

Nov. 1958 - Heat Transfer Reactor Experiment No. 3 began test at NRTS in Idaho. HTRE-3 was a development test assembly consisting of controls, shielding, direct cycle reactor and two modified J-47 jet engines with dual combustion chambers all mounted on railroad cars for moving out and into IET. The reactor was 51 inches diameter, 34.7 inches core length, 43.5 inches length overall with beryllium reflector and made up of 151 hex-shaped moderator cells of unclad hydrided zirconium with 3 inch bore for the fuel elements. The fuel element cartridges were concentric ring design, 19 stages each 1.5 inches long. The fuel elements sheet rings had 80Ni-20Cr alloy cladding over fully enriched uranium cores. Total weight of the powerplant assembly was twenty tons. During the shake-down test a power excursion melted some fuel elements. HTRE-3 was returned to IET hot shop for disassembly resulting in six months delay.

Dec. 1958 - Budgets for AEC and Defense Dept. cut back to $75 million each for fiscal 1960.

Dec. 1958 - Aviation Week broke the "news" that USSR had flight tested a nuclear-powered bomber and showed alleged photos of the Soviet plane. Newspaper editorials criticized the Administration for foot-dragging and Congressmen called for more funding to meet the Soviet threat. Defense Secretary McElroy was highly skeptical and Pres. Eisenhower said there was no reliable evidence of any kind and he was not changing the program to achieve early flight.

Dec. 1958 - GE-ANP employee paychecks and pay information set up on punch card system and printed by the IBM 650 Computer.

Feb. 1959 - Deputy Secretary of Defense Quarles briefed JCAE on ANP Program reiterating that the Program had to remain oriented toward development of a nuclear propulsion system rather than production of an aircraft until the material problems were solved. Representative Price countered with a challenge to the Eisenhower Administration to make an administrative decision and set a target date for achievement of nuclear flight. Convair was selected for airframe contractor.

April 1959 - At Rep. Price's suggestion, Deputy Secretary of Defense Quarles accompanied members of JCAE subcommittee to Evendale, Ohio along with government officials and congressmen for a first-hand look at the ANP Project. They toured the GE-ANP Evendale facilities and were briefed on direct cycle propulsion system developments. Price sensed that Quarles was influenced positively by what he saw.

May 7, 1959 - Deputy Secretary Quarles met AEC, Air Force and Navy officials to discuss the nuclear airplane. Both Maj. Gen. Keirn and Rep. Price felt that Quarles had decided to support a flight program. (Note: Although Quarles had been unwilling to support a crash program urged by DOD, he had protected ANP against those who might attempt to kill it, i.e. Defense Secretaries Wilson and McElroy).

May 8, 1959 - Quarles died of a heart attack. JCAE Open Hearings postponed indefinitely. Thomas Gates succeeded Quarles as Deputy Secretary of Defense, but felt unqualified to evaluate the ANP Program. That job fell to Dr. Herbert York, Director of Research and Engineering in Dept. of Defense, with supervision over all R & D projects, including ANP. York with Dr. A. Biehl prepared a new report on ANP.

June 7, 1959 - York gave his decision to JCAE and Rep. Price that ANP should be reoriented toward development of more advanced materials and that greater emphasis be placed on the indirect cycle. All target dates for nuclear flight were eliminated..

June 1959 - HTRE-3 with refueled core returned for test at IET facility.

July 1959 - Congressional hearing before Subcommittee on R & D of JCAE. First session on ANP Program.

July 1959 - Major redirection of reactor design from metallic to ceramic core.

Aug. 31, 1959 - Maj. Gen. Keirn retired from his dual role as Chief of ANP Office and Deputy Chief of Staff for Nuclear Systems in the Air Force. He was replaced by Brig. Gen. Irving Branch. D.R. Shoults left GE-ANP Program, transferring to GE New York Office. Sam Levine became Acting Manager..

Sept. 1959 - X-211 Turbojet Engine built in Evendale and tested in PUT cell using conventional fuel.

Sept. 1959 - DOD changed objectives of CAMAL mission to speed of Mach 0.8-0.9 at 35,00 feet with potential life of 1000 hours.

Oct. 1959 - HTRE-3 became critical and initially operated at 10 megawatts, meeting or exceeding all expectations.

Nov. 1959 - Six GE-ANP Scientists made first public review of the developments in rare earth metal technology (primarily yttrium) and refractory metal thermocouples at joint ASM-AEC Symposium in Chicago.

Nov. 1959 - GE-ANP personnel totaled 3650.

Dec. 1959 - HTRE-3 endurance tested at full power (31.8 megawatts) for 126 hours, 65 hours continuous operation with nuclear-power jet engines. Maximum fuel element temperature was 2030 °F. All components were in excellent condition. Both the reactor and engines performed in accordance with all design specifications. (Note: The HTRE-3 mobile test assembly is still intact at the Idaho Test Site).

May 1960 - Dave Shaw became Manager GE-ANP Department.

May 1960 - Flight Engine Test Facility (FET) completed at Idaho in a two acre building providing for operation of ground test prototype nuclear power plants mounted on test dollies as well as flight power plants mounted in an air frame mock-up. Also included in this facility were a hot shop with remote master-slave manipulators and a radioactive materials laboratory.

Nov. 1960 - Dr. Herbert York, Director of Defense R & D, concluded that the Nuclear Powered Flight Program did not measure up in competition for money and manpower. There was no longer a need for nuclear-powered aircraft in view of development of ICBM, short-range missiles, nuclear submarines, mid-air refueling and the many U.S. air bases in Europe.

Jan. 1961 - President Kennedy's Administration began. Robert McNamara appointed Secretary of Defense. The ANP Program received his personal attention and with the advice of Dr. Herbert York, he concluded that one of the budgetary items that could be trimmed at no cost to the nation's security was the nuclear airplane.

Mar. 1961 - AEC directed all contractors to discontinue any work related to nuclear-powered aircraft in view of President Kennedy's decision to cancel development program for nuclear military aircraft.

April 1961 - Ralph J. Cordiner elected President and CEO of General Electric Co. succeeding Robert Paxton.

May 1961 - National ANP Program terminated. In the 15 years of the ANP Program a total of $2 billion was spent on research and development. GE's Nuclear Materials & Propulsion Operation (NMPO) was established by AEC for continuation of basic research work on high temperature nuclear materials. Both Evendale Laboratories and Idaho Test Station continued work on future reactor projects including the 630A and 710 programs. General Manager of GE-NMPO was W. H. (Bill) Long. The GE-NMPO contract with the AEC continued until 1968.

REFERENCES AND SOURCES

Final Status Report of the Fairchild NEPA Project, NEPA-1830, T.A. Sims Jr., May 1951. Section 3, Technical History of the Project.

Hearing Before the Subcommittee on Research and Development of the Joint Committee on Atomic Energy, 86th Congress; First Session on Aircraft Nuclear Propulsion Program, July 23, 1959, Appendix A

APEX-901, GE-ANP Program Summary and References, June 28, 1962, G. Thornton, A. Rothstein, D. Culver

APEX-913, Metallic Fuel Element Materials, June 20, 1962, R.C. Lever, C.O. Tarr, R.K. Betts, C.P. Malone, M.L. Halterman. FPLD, GE Co.

ANPD News, Special Review Section, May 6, 1960, Vol. 4, No. 18

A & E Video, Investigative Report: "Nuclear Airplane" 1994

"Shooting Down the Nuclear Plane", W. Henry Lambright, Inter-University Case Program #104, Bobbs-Merrill Co., 1964

Acknowledgment: This document was made possible by the contributions and assistance in editing from Clay Brassfield, Bert Chandler, John Collins, Earl Funston, Coy Huffine, Tom Hunter, John Monday and Don Salyards.

BIBLIOGRAPHY

Aviation Week Magazine	1958, 1960, 1963
GE Aircraft Nuclear Propulsion Dept. Chronological History of NEPA and GE-ANP	Aug. 21, 1996
Nuclear Flight edited by Kenneth F. Gantz Duell, Sloan, and Pearce	1960
Race to Oblivion York, Herbert Frank Simon and Schuster Publishing Co.	1970
Hafstad, Lawrence R. - AEC Director	Speech 1949
Technical Report ANP Program Thornton, Gunnar	June 25, 1962

Preparing transport trailer for trial run with 597,000 lbs. of weights to test the roads and trailer.
Preparing transport trailer with railroad tracks to receive engine.
HTRE-2 being loaded onto transporter using cables and semitrucks.
Decontamination and decommissioning (D&D) efforts at the INEL were scheduled to include the complete dismantlement and burial of the HTRE-2 and -3 test assemblies. However, a review of environmental concerns related to INEL D&D projects in 1986, resulted in the proposal to preserve the HTRE test assemblies for its historical value. The Smithsonian Institution provided valuable guidance toward the preservation of those assemblies for use as a public display.
In preparation for the display of the test assemblies, the structure of each assembly were decontaminated and cleared of all loose materials. In addition, all pipe endings and other openings were plugged or otherwise sealed. The assemblies were then transported across the INEL approximately 35 miles to the EBR-I area. The transportation was accomplished by loading each assembly onto a special trailer and towing that unit on the existing INEL vehicle transport roadways.
The HTRE transporter was able to adjust its wheels to compensate for the centrifugal force and maintain the level of its bed during turns or corners.
INEL
TAN
CTF
PBF
WRRTF
TREAT
EBR-II
NRF
CPP
WERF
CFA
ARA

David M. Carpenter

A Maine native and graduate of Northeastern University, Dave has worked for General Electric Aircraft Engines since 1963. Presently he is plant historian having retired as Manager of Spare Parts Sales for the T-700 Turboshaft Helicopter engine that powers the Black Hawk and Apache Aircraft.

This is Dave's fifth book about General Electric Riverworks history. His first book, "Flame Powered," covered the development of America's first jet engine here at the Riverworks and the aircraft it powered. His second book was about the Riverworks plant itself. Entitled, "Riverworks", City within a City. The third was "Jet Deck" and the fourth was called, "Thunderbug" about America's first jet propelled cruise missile the JB-1A in 1944.

General Electric Co.
1000 Western Avenue
Building 232
Lynn, MA 01910

- The Last Word -

Between 1956 (start of HTRE-1 testing) and March 1961 when the program was terminated only the HTRE-1, the HTRE-2 and the HTRE-3 had been operated at their design power. None of these designs were suitable to power the flight of suitable aircraft. They were "heat transfer reactor experiments".

However a breakthrough was on the horizon. This came in the form of HTRE-2 testing involving a ceramic (BeO) impregnated with Uranium Oxide. "The ceramic fuel element". This testing showed a temperature capability in excess of 2700DEG f with bulk air temperature in excess of 2000DEG f. Just what the doctor ordered. The design effort of the last three years of the program was directed to the ceramic core design and the XMA-1 power plant. This was a true aircraft power plant capable of aircraft installation and propulsion.

Heat Transfer Reactor Experiment (HTRE-2)
The 600,000 plus lbs Nuclear Turbojet Reactor Engine Assembly being moved to its final resting place in Arco Idaho, 1985.

ISBN 0-9633387-9-X
50995
9 780963 338792